Transfer of Technology

István Gazda

Transfer of Technology

by István Gazda

1996

Kluwer Law International
The Hague-London-Boston

Kluwer Law International
P.O. Box 85889
2508 CN The Hague/The Netherlands
Tel. +31 70 308 1561
Fax +31 70 308 1515

Sold and distributed in the U.S.A. and Canada
by Kluwer Law International
101 Philip Drive, Norwell, MA 02061,
U.S.A.

In all other countries, sold and distributed
by Kluwer Law International
P.O. Box 322, 3300 AH Dordrecht,
The Netherlands

Cover design: Robert Vulkers (bNO)

ISBN 90 4110 044X

The original work was published under the title: *A technológia transzfer* by the Közgazdasági és Jogi Könyvkiadó, Budapest, 1993.

The present edition is a translated and somewhat revised version of the Hungarian issue.

Translated by the author.

Revised by Dr. Elemér Nagy.

Preface

Serious efforts are to be made to adapt developed foreign technology in the period of transition towards the market economy. Although this activity is supported by the incoming foreign capital and the formation of joint venture companies, there are still some sectors of the economy where the national property remains dominant and where the adaptation of developed (foreign) techniques or technology is unavoidable. It is also necessary to give some practical hints concerning the trade with developing countries and the transfer of technology to them.

The aim of the present book is to promote the knowledge of the transfer methods, to get acquainted with the concomitant legal techniques, and at the same time to help the entrepreneur to avoid unforeseeable difficulties. The treatment of the issues is not separated into distinct import and export chapters, the emphasis being laid on the main general features of the different subjects.

In *Chapter 1* we clarify the basic concepts. It concentrates relatively less on those basic concepts which are extensively discussed in the literature (e.g. patents, trade marks), and more on e.g. know-how, the literature on which is relatively poor.

Chapter 2 wants to show, how the transfer of technology takes place, while *Chapter 3* emphasises those legal issues which were the subject of the Third International Private Law Conference (Stockholm, 10–14 June 1991, dealing exclusively with the transfer of technology).

Chapter 4 acquaints the reader with the contracting practices. It deals with each contractual issue separately and vertically. That means that *in lieu* of breaking the subject up into the questions of concluding, fulfilling and breaching of contract, it seeks to handle the main components of each kind of contractual issue at the same time.

Chapter 5 deserves a special interest as it deals with transfers connected to the foundation of a company. The attention of the reader is however drawn to the fact, that this chapter deals with the formation of a company only to the extent necessary in the context of technology transfers. Very rich literature is namely available about the formation procedure of a company itself.

The author hopes that this book – which is the first to summarise the transfer of technology in the Hungarian literature – shall contribute not only to the foundation of Hungarian literature on the subject, but shall also give practical knowledge concerning the fulfilment of this important task.

Preface to the English Edition

The present book has been translated from Hungarian into English and it has been revised to a certain extent, more having been written about Swiss law and less about Hungarian. It was thought, however, that the total exclusion of Hungarian legislation would be wrong, since the reader may be interested in how these concepts are introduced in a country in the process of development of a market economy.

A few words on the technical terms. The reader shall immediately see that the terms seller, grantor, transferor, lessor, etc. are used to express the same person (natural or legal), who is giving and/or transmitting something, while the terms buyer, grantee, transferee, lessee, etc. are used to denote the acceptor of a delivery or a service. These are synonymous expressions, as are agreement, deed and contract, as well as construction, erection and mounting, assembling, etc.

The reader will find a lot of cross-references in the book and even a few repetitions. The reason for the latter is that should the reader open the book at any place, he should get a full review of the matter concerned.

The author has worked a lot on the theoretical backgrounds of the subject and is at the same time a man of the practice, who has worked for a long time in the field of the transfer of technology, and he hopes that the reader shall find both the theoretical and the practical knowledge in the present book.

Budapest, August 1994.

I. Gazda

Table of Contents

1 Basic Concepts

1. TECHNOLOGY

According to Webster[1] the meaning of the word *technology* is as follows: 'the science or study of the practical or industrial art.' If we take this meaning of the word technology into consideration, it incorporates both the technical sciences and the practical arts, knowledge in a broad sense. It means that besides the 'classical' technical sciences (e.g. mechanics, electricity, etc.) it includes such technical areas as for instance the technical aspects of biotechnology, biology, horticulture, zoology, etc., all of them in a very broad sense.

Taking now the practical sciences into consideration, it includes such knowledge and experience as the financial, economical and commercial ones (e.g. how to found and run a hotel, a petrol station, a shop or a bank). In the following we shall understand under 'technology' both the technical and practical sciences in a broad sense. At the same time it should be remarked that only a part of technology – as defined above – is patentable, while a very large part of it is not. The concept of *know-how*, however, covers the non-patentable part of technology as well.

2. TRANSFER

According to Webster[2] the word *transfer* has a lot of meanings. Among others it means: 'to convey, carry, remove or send from one person, place or position to another.' In law – also according to Webster – it also means: '(a) transferring of a title, right, or property from one person to another, (b) the document effecting this.'

In the following we understand and use the word 'transfer' in this very broad sense, and it shall mean on the one hand the introduction of technology to another person, and on the other hand the transfer of a title or right, as well as the *legal effects* of both.

The intricate concept of transfer may be more easily treated through an appropriate classification, the main types being:

1. Webster: *New Twentieth Century Dictionary*, 2nd edition, Simon and Schuster, 1983, p. 1872.
2. Webster, *ibid.*, p. 1938.

- the unpackaged transfer and
- the packaged transfer,

allowing of course for intermediate cases as well.

2.1. Unpackaged transfer

By an *unpackaged transfer* we mean that only the title, the right in itself, the essence of technology (i.e. the technical or other practical knowledge) together with drawings and/or descriptions is transferred, with a so-called basic engineering at most (*see* paragraph 7.1 of this chapter). Should e.g. a formula be transferred in the framework of an unpackaged transfer, one might suppose that the recipient disposes of specialists in the technical field in question who are able to utilise the transferred technology.

Such types of transfer usually occur between partners who are at the same technical level.

2.2. Packaged transfer

The packaged transfer means that besides granting the right of use and describing the technology to a minimal extent, a good deal of other information is also transferred to the recipient. The packaged transfer may imply – depending on its complexity – the following:

- a basic engineering (although this might be a part of the unpackaged transfer too),
- a detailed engineering (*see* paragraph 7.2 of this chapter),
- delivery of machinery,
- delivery of a complete plant,
- the direction and/or supervision of the construction and/or erection works or the partial and/or full execution thereof,
- the supervision or execution of the putting into operation and the commissioning of the plant,
- a so-called turnkey job.

This type of transfer is usual between unequal partners, i.e. partners who are not at the same technical level.

3. INVENTIONS, PATENTS[3]

Patents are exclusive rights to technical inventions, limited however territorially (e.g. to one country or group of countries) and also in time (generally for a period of 15–20 years, in some cases for less).

The patent protection has to be asked for, that is, one has to file an application for it. There are formal requirements which must be observed before the application is investigated. In respect of its contents strict rules are laid down.

First of all: those inventions for which one applies for a patent, have to be *new*. By this is meant that the invention may not be described prior to the application in any book, article or patent specification – either at home or abroad – nor may it be in public use prior to the filing of the application. An invention which has been described in a publication prior to the application generally loses its novelty except in some countries, e.g. the United States, where a patent can be applied for up to one year following the publication by the inventor.

In almost all countries it is a further requirement that the new invention has to represent an advance or progress compared to the existing art, and it has to have a certain *inventive level* (in German: *Erfindungshöhe*). It must also have a technical effect and reproduction must be possible.

A patent application has to contain a description of the invention, its practical applicability (if necessary with examples and drawings), while the features to be protected have to be summed up in one or more patent claims.

There are still some countries where chemical materials, medicines and food products are excluded from patentability (up to last year this was the case in Hungary), but the processes (the technology) leading to their production may be patented. For new plants and animals there is a limited number of countries (including Hungary) where one may apply for a patent.

The application is then examined. Should it be found that it is in order in formal respects, that it is new and in most of the cases also represents progress, the patent is granted, leading to a patent protection.

In many countries (including Hungary) the application is disclosed or published irrespective of its granting after the lapse of a certain time. In other countries only granted patents are published. Where the application is disclosed with a provisional text prior to its grant, it will be republished in its final form.

In most countries the patent has to be put into practice after its granting, that is, it should be used or the use should be licensed to others. The lack of use may lead to the granting of a compulsory licence (*see* paragraph 12.3 of this chapter) after 3–4 years.

3. The main rules for patents are more or less the same in most countries. The Hungarian patent law is based on Act II of 1969 as modified by the Law-Decree No 5 of 1983, and Act VII of 1994 which entered into force on 1 July 1994. A new Patent Act has been enacted recently (Act XXXIII of 1995) which shall enter into force on 1 January 1996 and shall be valid for applications filed thereafter. This does however not alter the principles discussed below.

Previously patent applications had to be filed in each country separately, but nowadays there is the possibility to achieve through a single application, to be filed at the European Patent Office in Munich, a patent for those countries which have signed the European Patent Treaty. Since this treaty is an open one, it is possible to achieve a patent for the mentioned countries by filing a single application at this Office, if one's country is not a signatory to the mentioned treaty but is a member of the Paris Union Treaty.

There is also a treaty which allows for a common application worldwide, although not all countries are as yet party to it. This is the Patent Cooperation Treaty (PCT). The corresponding patent applications have to be filed in Geneva. It offers some advantages insofar as one may apply at a single office, but the applications have to translated later into the language of the country where the application is to be treated according to the home patent law. The European Union is one of the members to this treaty.

The Paris Union Treaty, one of the pioneering international treaties, now with more than 100 member nations, stipulates that following application in any one of the member countries one is allowed to file the same patent application in any other member country within one year of the first application. In this case the application is made with *retrospectivity*, i.e. the novelty of the invention is to be regarded in respect of the date when it was filed in the first member country.

In most of the countries certain fees or duties must be paid (mostly yearly) to maintain the legal force of a granted patent. Should these not be paid – after the lapse of a certain period of grace – the patent protection ceases to exist.

The owner of the patent (the patentee) has the *sole right* to put the patent into practice, to use it and to sell the products thus produced, or to allow the use of the patent by others, i.e. to grant a licence. Should the patent cease to exist, the invention may be used by anyone free of charge.

The illegal use of a patent forms an infringement. An infringement is committed if anyone produces or sells a product corresponding to an invention for which a valid patent has been granted. The infringement generally calls for the compensation of damages, but may involve a penal liability too.

4. UTILITY MODEL

The utility model is an exclusive right similar to the patent system. It concerns the grant of an exclusive right to the shape of an article, its construction or the arrangement of its parts. A utility model protection exists in more than 60 countries of the world, but not everywhere. Its importance may however increase in the future.

The utility model is relatively new in Hungary: Act XXXVIII of 1991 introduced it with legal force as of 1 January 1992. It is known generally in international literature as a '*small patent*,' because products of minor value (e.g. a new type of table lamp) are protectable in such a way. The novelty is a must here too, but only a relative one. In lieu of progress the requirement of an inventive step exists, meaning that the subject of an utility model should not be an obvious one considering the state of the art of those persons skilled in the field. One cannot protect processes through a utility model, nor any horticultural or animal species, nor the aesthetic appearance of an article.

The utility model gives – in accordance with its smaller importance – a shorter period of protection than do patents. This protection varies in the world (it is 10 years in Hungary), and for the maintenance of protection fees have be paid (in most cases yearly) which are lower than those for patents (in Hungary they are half of these).

The procedures for obtaining a utility model are similar to those for the obtention of patents.

5. TECHNICAL KNOW-HOW

Know-how has no generally accepted definition. It is quite a ramified concept which is not uniform in various countries. The literature does not deal too much with this concept. There are only a few books and a number of articles dealing with this matter.[4]

We shall adhere in the following to the IAPIP definition adopted at the Congress of Melbourne (1974), which states:

> 'Know how is knowledge and experience of a technical, commercial, administrative, financial or other nature, which is practically applicable in the operation of an enterprise or the practice of a profession.'

This definition does not cover technical know-how only, but extends to non-technical know-how too. For practical purposes we shall deal with them under separate headings.

Know-how is protected in Hungary by the Hungarian *Civil Code* of 1978, which fact is more or less unique. According to paragraphs (3) and (4) of § 86 of the Civil Code:

> '(3) The law protects also intellectual creations which are not protected under special regulations, if they may widely be used in society and have not yet come into the public domain.
> (4) Protection is due to persons also with respect to their economic, technical and organisational knowledge and experience of pecuniary value. The

4. See e.g. Gazda, *Az ipari know-how* (The industrial know-how), Budapest, 1969 (in Hungarian only); Gazda/Kövesdi/Vida, *Találmányok, szabadalmak* (Inventions, patents), Budapest, 1985, p. 123 ff. (in Hungarian only). Other references are given in the bibliography to this book.

> commencement and the substance of protection are defined in separate regulations.'

This protection is also strengthened by the Unfair Market Behaviour Act (LXXXVI of 1990), which states:

> 'It is prohibited to acquire a trade secret in an unfair way, or to publish or use the same without authorisation.'

In respect of what forms a trade secret the law says the following:

> '... a trade secret is all such solutions, or data which are connected with the economic activity and to the secrecy of which there exists a reasonable interest of the entitled.'

It is clear that these provisions of law do not only cover technical know-how, but also non-technical know-how. Technical know-how is hence knowledge and experience which may be used in practice and to the secrecy of which an interest is attached.

Know-how, including technical know-how, is protected in other countries indirectly through the various unfair competition laws. It must be remarked, however, that the safest protection is achieved by the careful formulation of the contracts dealing with the transfer of know-how.

5.1. Constructional know-how

The nature of *constructional know-how* can best be illustrated by an example. It is apparently not a difficult job to devise equipment with a long steel conveyor belt. If, however, the operational conditions are such that this belt has to pass alternately through a heating zone at 250°C and a cooling zone at room temperature, the implementation may not be easy at all. The difficulty increases if the full length of the belt is, say, 100 meters and the width 2 meters. Hitherto only a single machine building company had adequately solved this problem through small refinements at the drive. These tricks, which eliminated the break-downs common at other companies, may be regarded as constructional know-how.

Similar know-how is present in most branches of industry. For example, a good deal of constructional know-how is to be found in the ventilation and drying industry. The types of air filters to be used, how the ventilators have to be dimensioned, what should be the route of the air, how to choose the material of the air-ducts, the mixing ratio of fresh to used air, that of saturated used air to fresh air, the shape of the drying routes – all make up a knowledge or know-how based on practical experience. How a given job is to be performed represents another kind of knowledge or know-how.

5.2. Designers' know-how

The concept of *designers' know-how* is broader than that of constructional know-how. Its most typical example is the special knowledge necessary for the lay-out of a production line, a plant or a workshop. The diverse technically possible solutions can show quite considerable differences in investment and operational costs, and therefore may decisively influence the profitability, marketability, etc. of the product. The good utilisation of space decreases the building costs and allows for shorter and more economic energy lines. An open-air location may reduce costs further. Special skills and knowledge are necessary in the designing phase.

Nowadays the finding of the most economic environmental protection method is very important to reduce the costs. This may affect not only the cost and input relations of a single plant, but also of a whole branch of industry, as was the case for aluminium smelting. The necessity of the elimination of fluoride-containing gases led to a series of technical modifications in the preceding production phases as well.

We have to emphasise that to find suitable solutions ideas and inventive steps are necessary which come very close to patentable inventive steps. Such knowledge can only be achieved by long experience or lengthy experimentation. Among designers' know-how the most valuable is that which finds an optimal solution without long and expensive experimentation. (How much experience is necessary, e.g. for the preparation of a pipeline design of a plant or its electrical wiring design, can only be appreciated by such persons who were present in the designing phase and who heard the discussions connected therewith.)

It makes no difference at all whether the know-how is in the form of computer software which contains and makes available in a user-friendly form the knowledge and experience of the specialist(s). The software itself may, of course, be valuable irrespective of its contents, and it is protectable by copyright. This issue however is to be settled between the know-how transferor and the software maker, and not between the transferor and transferee.

Designers' know-how is necessary not only to find the most practical lay-out, but also for the preparation of the detailed engineering.

One may bring up many other examples of designers' know-how, since it plays a role in almost all technical fields. It must be emphasised, finally, that there is no strict border-line between constructional know-how and designers' know-how: in many instances they are interwoven. This is the reason why they can hardly be separated from the concept of engineering (*see* paragraph 7 of this chapter).

5.3. Process (technological) know-how

Another type of know-how, *process* or *technological know-how*, is the only form of know-how known to the general public. A large part of know-how may fall into this category.

When dealing with technological know-how, the first issue is to determine its relationship to a patent. Some of the know-how relating to a process improvement does not represent a substantial development, or may be found in literature, albeit in another connection or with a negative conclusion. It is questionable whether such improvements are patentable, and it is therefore advantageous to avoid engaging in a dubious patenting procedure. Really valuable know-how is certainly patentable, but it is not advised to attempt it. For example, when an end product does not reveal how it has been made and whether or not a certain patented know-how has been used, an infringement suit is hopeless since there is no possibility of proving the infringement. Taking out such a patent therefore can imply offering one's know-how to competitors free of charge, sparing them research and development costs. In such cases the inventor who wants to make his invention a lucrative proposition does well to keep it secret and utilise it as know-how.

Process know-how may be divided in principle into two basic groups, a gradual transition between them being however possible. One of the basic groups is the *individual process technique*. For instance, in chemical processes an inserted new operational step, or the elaboration or modification of the pressure, temperature, the catalyst used, the flow conditions, etc., or several such techniques put together, form know-how of this kind. Such details may be connected to patents too, as a not separately patented and hence not precisely described portion of a patent.

The other basic type is that of the *process as a whole*. An example was the practical realisation of high pressure polyethylene synthesis some 25 years ago. The know-how included the application of a pressure of 1000 atmospheres, which required the use of special technologies determined by the purity of the raw materials, the catalysts to be used and nearly all steps of the process.

In between the two basic types of process know-how is the transitional type, i.e. where not the process as a whole but some of its individual, well-defined and distinguishable steps form the process know-how. A good example is the production technology of some types of non-woven textiles (so-called 'vlies' materials). Depending on the type of non-woven material to be produced and its characteristics, the extent of privately owned know-how differs. The production techniques for some types of non-woven materials are within the public domain, while others may only be produced with the help of hard to get know-how.

5.4. Operational know-how

This type of know-how differs from the process know-how in that it concerns those typically small techniques which are necessary for the continuous operation of a process in cases where the production is not yet fully automated. These techniques do not touch the essence of the process (technology) but only the operation itself, i.e. the elimination of operational problems or trouble-shooting if a problem arises. Such knowledge can only be gained through experience. A lack of this kind of knowledge or know-how may lead to considerable damage. Similarly, knowledge and experience which allows production to take place under optimal conditions or with a low level of waste may represent operational know-how. Such operational knowledge may be extremely valuable in any branch of industry.

This kind of know-how may be acquired only through training and practical experience, and not by reading, so the acquisition of such knowledge is promoted by training agreements in connection with a know-how agreement which forms part of a transfer (*see* paragraph 2.2 of Chapter 2).

5.5. Compositional (formula) know-how

This form of know-how differs from those discussed above, although it often gives a full knowledge together with the process know-how. The essence of a technology often lies in the formula itself, but in other cases rather in the process steps, and sometimes in both.

Formulas which are publicly known or which are an easily analyzable mixture, cannot be regarded as know-how. In cases where the formula is not publicly known and the composition of a product cannot be determined by simple means, a compositional or formula know-how exists. This is often the case where the components are subjected to a chemical transformation which is difficult to trace through an analysis. An example is a formula for the production of *plastics*. It is not usual to patent such a formula. Another typical example of compositional know-how, which is best kept secret, is *rubber mixtures* (the species of raw material and the types of additives, accelerators and softeners) as well as the catalysts used (their grain-size distribution and some characteristics of their base material).

5.6. Other technical know-hows

One may find various other forms of know-how besides those discussed above. Without attempting to be exhaustive, we mention the following:

- in case of the erection of complicated equipment, devices or machinery, we may speak of *erection know-how*, which means the relevant knowledge of specially trained experts;
- we may speak of *reparation know-how* when the special expertise renders the quick and precise elimination of a defect possible,
- some *computer programs* which are under the protection of the copyright law but which go beyond the so-called routine software (e.g. some production optimisation programs and equipment dimensioning programs, as well as so-called supervision programs of computer systems).

5.7. Complex know-how

At the present level of technical development it is often the case that in order to achieve an aim a lot of know-how, so to speak a chain of know-how, is necessary. The elements of this chain (the chain-links) may naturally form individual patents. This type of know-how may be called *complex know-how*. The value of each part thereof depends on the case.

The concept of *complex know-how* characterises cases where the *proceeds of the full know-how are larger than the simple sum of the proceeds arising from the parts thereof.*

A substantial part of the more valuable know-how today belongs in this category. Nearly all complicated processes and those which consist of more than one step (multistep processes) – not to mention those which require a new plant – are coupled to complex know-how. The individual part solutions are in such cases insufficient, as the desired result can only be reached through the organic chain thereof.

A typical example of complex know-how is ammonia production, for which several different techniques exist. The whole procedure is a multistep one (it contains at least 6–7 important steps) and each has its specific problems. If a modification takes place in one production phase, the advantage this brings may be lost in the other non-modified steps. The harmonisation of these steps, the elaboration of the optimal energy economy method, the introduction of automation, the finding of the most economic lay-out of the plant, etc., in their totality surpass by far the framework of simple engineering or routine and can only be executed through the possession of complex know-how.

An individual company does not necessarily dispose of the necessary complex know-how and hence *ad hoc* associations are formed for the solution of such problems. The combined knowledge of the companies which form the association then constitutes the complex know-how.

5.8. System know-how

Under *system know-how* we understand knowledge of a complex unit for satisfying a demand consisting of different types of parts, which are complementary but suitable only in their totality for satisfying the specific demand. The unit may serve for industrial or agricultural production, but may also meet e.g. an educational or even a public health demand.

The component parts may be production units or the buildings which house them, or basic or auxiliary materials for production purposes, but may also be organisational, distributional, medical or other types of activity.

Typical examples of systems know-how can be found in agriculture and animal husbandry.

A system for the production of maize (Indian corn) may cover the entire process starting from seed, soil cultivation and the necessary machinery, harvesting, storage of the produced maize, and eventually its utilisation or further processing.

A livestock system may include breeding animals (as well as the development of superior breeds, the means for and technology of breeding), buildings, preparation and storage of foodstuffs, feeding equipment, the unavoidable veterinary assistance, and even how the animals should be processed (slaughterhouses, meat works). It may also include the production of veterinary preparations and their formulae.

In other systems the non-technical elements may play a greater role. So e.g. a natural gas filling and selling system may comprise, besides the filling stations themselves and the filling technology, a plant and the technology for the production of bottles, a method and the equipment for monitoring the produced bottles, but mainly the organisation of the distribution of the bottles, the collection of empty bottles and some other complementary activities. The service and training systems may be similarly complex.

These systems require another type of complex knowledge than the complex know-how discussed above, as a part of the complexity is non-technical. System know-how differs mainly in this respect from complex know-how.

6. NON-TECHNICAL KNOW-HOW

Both the IAPIP definition (*see* paragraph 5 above) and the provisions of Hungarian national law acknowledge *non-technical know-how*. The definition of the IAPIP refers also to 'commercial, administrative, financial or other' knowledge and experience. Paragraph (4) of § 86 of the Hungarian Civil Code mentions 'economic, ... *organisational* knowledge and experience of pecuniary value.' These two definitions of non-technical know-how are practically identical. For reaching a certain result it is of secondary importance whether technical or non-technical means are used. If we wish to create a service system, the *organisational* knowledge and experience shall probably be preponderant. In the setting up of a commercial network, *commercial* and *financial* knowledge and experience shall dominate; in the case of a banking system financial knowledge shall be important. The operation of a non-technical establishment such as a special school or a hospital requires *operational* know-how. Another example of such a non-technical establishment is a hotel, the starting up and later the continuous operation of which needs commercial, financial, organisational, operational and of course *specific* hotel skills (catering, tourism, etc.). The commercial ('market') knowledge is a decisive factor in securing the customers of the hotel. Such a complex unit requires however some technical knowledge too (the operation of the boiler house, the kitchen and other technical installations).

The domain of non-technical know-how therefore includes the following:

- organisational,
- commercial,
- financial,
- operational,
- non-technical professional (branch specific),
- complex non-technical know-how.

The concepts of the first four types of know-how are self-explanatory and these can be quite similar in very different activities. By contrast, 'non-technical professional know-how' means that special knowledge which is derived from experience in various non-technical sectors which are very different from each other.

How diverse non-technical know-how may be can be shown with the example of a hotel. A packaged know-how transfer relating to a hotel may – among other things – include the following:

- feasibility study of hotel building and funding;
- cooperation in the conceptual planning of the establishment (location, visitors catered for, category, size, services);
- assistance in the architectural and interior design, the drawing up of a detailed list of investments (furniture, machinery, etc.), and eventual assistance in their acquisition;
- proposals for developing international contacts (affiliation to hotel chains, agreements with tour operators, etc.);
- assistance in the training of the employees and in the designing of their uniforms;
- assistance in establishing an appropriate system of accounting;
- assistance in the advertising activity, including participation in the promotion activities of the know-how grantor;
- handing over of other knowledge and advice in order to maintain the service at the level and to achieve the atmosphere in the hotel which is specific to the grantor;

- handing over of the knowledge necessary for the operation of the hotel administration,
- handing over of the knowledge necessary for the operation of the technical installations and equipment of the hotel.

The main difference between technical and non-technical know-how is that while the former resembles patentable inventions, an important feature of non-technical know-how is that *its individual elements do not constitute any invention,* hence no patent protection can be secured for them.

6.1. Organisational know-how

This is the most general form of non-technical know-how. It includes among others the organisation of a plant (factory) or commercial establishment, including complex institutions like the above-mentioned hotel. In general the specific organisational knowledge is accompanied by technical knowledge. In the case that an expertise company is asked to optimise the operation of an institution, it can undertake this job seriously only if it can determine by a full technical and economic survey of the plant or institution whether, how and where the chain of technology must also be modified.

6.2. Commercial know-how

If one has either a specialised commercial enterprise or the commercial department of a factory or company in mind, it becomes at once clear that a great deal of experience and knowledge is necessary for setting them up or putting them in order if they malfunction. The administrative steps required, the circulars to be issued, the conduct of market research and the building up of a clientele, require know-how. A stable clientele, constituting the goodwill of the enterprise, is an important trade asset of not inconsiderable value. The keeping of records is the duty of any company, but an essential element of commercial know-how may be the choice of data to be recorded and software to be used.

Another important branch of commercial activity is promotion (advertising) activity. Although we will not discuss this activity in detail, the presence of a know-how component is clear.

6.3. Financial know-how

Financial know-how comprises manifold, complex knowledge and experience. This concept covers the knowledge necessary for the elaboration of the financial policy of industrial and commercial enterprises, but also the special experience needed for the setting up and operation of a bank or a brokerage company.

Financial know-how is first needed when one seeks the optimal operation form of an industrial, commercial or service enterprise, and later when deciding what financial business politics are necessary for the most favourable operation (e.g. the amount of total credit and the mix of credits maturing on short-, medium- and long terms).

There is no question that for the establishing of a bank a multiple, extensive financial know-how is needed. Here we have not only to take into consideration the above-mentioned aspects, but also in which environment the establishment takes place (e.g. in a yet underdeveloped area, the development of which is however to be expected – because it is near or between traffic hubs – or in a strongly industrialised area). Over and above this, consideration must also given to what should be the duty of the given bank, what its capital policies should be, and what type of source, credit and business policy it is to have. All these need very aggregated knowledge and practical experience, only the main theoretical elements of which can be learned at universities or taken from the literature. The person or group which undertakes to found a bank has to have comprehensive financial know-how.

6.4. Non-technical operational know-how

The operation of non-technical enterprises – such as a commercial enterprise, a banking institute, a business network, an advertising enterprise, a petrol station chain, a hotel chain etc. – presupposes and demands experience and knowledge, i.e. know-how. These together could be called non-technical operational know-how.

6.5. Non-technical professional know-how

This concept comprises the specific knowledge of a branch or of a professional sector. For example, the professional skill needed for a service activity, let's say of a TV company, is specific for the branch and differs from the specific skills found in other branches.

It is obvious that a washing machine and a car repair shop are both involved in a service activity, and it is possible that they use some know-how which is common to any kind of repair shop, but there is also specific know-how involved which is characteristic for the washing machine or the car repair shop. The concept of non-technical professional know-how applies to this latter type.

6.6. Complex non-technical know-how

Above we already gave the example of the hotel, but many other enterprises have equally complex activities. The know-how which is connected to such combined activities shall be called complex, non-technical know-how.

7. ENGINEERING

According to Webster[5] the word *engineering* means:

> 'the planning, designing, construction, or management of machinery, roads, bridges, buildings, fortifications, waterways, etc.; science, profession, or work of an engineer.'

According to other definitions it means *inter alia*:

> 'an engineer's work or science, machinery, techniques, machine designing, mechanisation.'

Engineering has hence a lot of meanings, but it could be simply called a kind of technical work involving some designing activity.

Its economic-legal implications are however broader. The activity of technical designing could be called engineering, independently of the question of liability, but so too could a technical service activity which contains designing elements and which is connected over and above the professional warranty to a pre-determined technical liability for the result. Indeed, the existence of liability for a design is tied up with the definition of engineering:

- The *designing activity* is the 'classical' engineering, which is a far-reaching activity (as shall be seen later on), only part of which is subject to a 'professional' warranty (e.g. the so-called detailed engineering).
- The non-designing *technical services* in general do not entail liability for the result (with eventual exceptions); such services belonging generally to the concept of 'technical assistance,' which is *not* identical with the engineering.
- A *complex designing activity* is often coupled to one or more 'technical assistance' activities (the liability for the result can be assured and undertaken thereby), which merge into the designing activity and become a supplementary part thereof (this is the reason why the concept of engineering ought to be extended).
- The *technical advisory activity* is so to speak a category in itself and involves a limited liability.

5. Webster: *New Twentieth Century Dictionary*, 2nd edition, Simon and Schuster, 1983, p. 602.

The clarification of the concept of engineering is not an academic question, but in connection with contracts and agreements of practical importance. Should one use the concept without determining its meaning, it may cause misunderstanding in the fulfilment and interpretation of the contract and in the determination of the scope of liability.

The present practice is that while engineering contracts are formulated either on an *entrepreneurial* basis (in German: '*Werkvertrag*') or on a *commission* basis, for technical assistance contracts and advisory activities the commission basis is preferred. The entrepreneurial or commission basis of a contract influences the price, the conditions of payment, and some other relevant terms of the agreement.

7.1. The feasibility study

The feasibility study investigates a project in broad lines, proposes technical solutions, and deals with the questions of financing and marketing. Although it is not a really technical plan but a preliminary study which comprises economic elements too, it comes nearest to the concept of engineering. The investment program has a similar character which – if the aim is the realisation of an industrial establishment – may also come under the umbrella of the concept of engineering. These activities can be regarded as engineering of a commissionary character.

7.2. Basic engineering

The most usual form of engineering is basic engineering, following which the details of a plan can be elaborated. Such basic engineering is necessary for the industrial realisation of technical solutions, and the technical data for the given capacity may be deduced from it. It includes e.g. a general layout, the specifications and data sheets of the required equipment (but not their detailed drawings), the type and quantity of energy supplies (but not a detailed electric, instrument or piping network), a general description of the needed raw materials needed, and methods for quality control.

It is not usual to hand over with the basic engineering the drawings of the buildings to be built based on the know-how to be transferred; rather, only those initial data on the basis of which the transferee may prepare its own building projects or designs are delivered. The question whether the initial data were adhered to by the transferee is usually controlled by the party giving the basic engineering.

In general the basic engineering does not include data for auxiliary equipment. If e.g. steam is needed for the operation of the equipment, the transferor lists in the frame of the basic engineering what quantity and quality of steam is needed, but it is not generally his job to supply designs of or data for the boiler plant. Should a crane be needed for the operation, the transferor simply remarks this fact and it is the transferee's task to choose and install the right crane.

This is the typical engineering which is generally delivered by the transferor of the know-how, or in case of patent licences by the licensor. It is clear from the aforementioned that basic engineering is an important form of the transfer of technology.

It has to be remarked that the liability for the result is justified mainly because the basic engineering embodies the essence of the technical solution.

In case of a proper basic engineering, the detailed engineering can be made either by the transferee or by another designer.

7.3. Detailed engineering

The detailed engineering is connected to the different phases of the realisation of a project. Depending on which detail of the execution it relates to, it is usual to speak of specific fields, e.g. mechanical, chemical, instrumental or electrical engineering.

As already mentioned, the detailed designs may be produced by the transferee if supported by a sufficient basic engineering. Experience shows, however, that the basic designs do not always contain all the relevant data in sufficient detail, so for an unskilled transferee the right solution may be to order the detailed engineering as well from the transferor or from another competent partner.

7.4. Consulting engineering

This concept is quite wide. According to some opinions it can hardly be defined[6] but it may basically be regarded as an intellectual service, executed by a generally stable group of experts of varied skills (often a legal person). The given group of experts has to be thoroughly familiar with the current ideas and developments, must be capable of checking their validity for the project in question, and should have a clear insight in the economic development of the field, particularly regarding the realisation of industrial establishments.

The consultant may also be a single person instead of a group of experts.

The activity may be realised in the form of written studies (e.g. the feasibility study mentioned in paragraph 7.1 of this chapter), or by giving an oral or written opinion relating to technical-economic questions. It may also concern

6. This was the opinion communicated by Dr. Alfred Hubert (from Belgium) during the AIPPI-MIE Conference in Budapest in 1970 and published in the relevant Conference papers (AIPPI-MIE Conference Budapest, 1970, p. 217 ff).

technical control, including technical coordination. To sum up: the consulting engineering activity may be a technical-economic consulting activity, an activity where expertise is applied in an advisory capacity, or a controlling and/or coordination activity.

7.5. The difference between know-how and engineering

The problem often arises that the documentation of the know-how is mixed with the technical documentation corresponding to the basic engineering, and sometimes with the plans of the detailed engineering too. So the transferee does not know where the one ends and the other one starts. This is important in connection with the secrecy obligation generally to be undertaken by the transferee in the framework of the transfer (see paragraph 2 of Chapter 3). It is therefore advisable to lay down in the agreement the scope and limits of the know-how to be transferred and even to mark on the documents themselves to which category they belong. This is not only necessary because of the secrecy obligation, but also to avoid possible illicit use.

Although the operating manual which is delivered together with a machine (i.e. the description of its use and trouble-shooting) forms an inseparable part of the machine, its unlimited use may be hardly forbidden, although it may contain know-how. At the same time the delivery of a plant unit consisting of a series of machines with the corresponding production process comprises in general know-how which deserves protection.

The transition between engineering and know-how is often indistinct, since engineering may be regarded as an *applied technical creation* or applied know-how.

The question whether know-how is accompanied by basic engineering or not depends also on the nature of the know-how. There are cases where it is unusual and unnecessary to supply engineering with the know-how, e.g. if the know-how consists of a formula (see paragraph 5.5 of this chapter) or if it may be executed on the existing equipment without their modification, or if it consists of practical techniques which can be put into practice in a routine way.

While basic engineering can often be regarded as an accessory to know-how, detailed engineering does not have this character. This is shown by the usage of giving it a separate price tag. The main reason for this practice is that the basic design has to be adapted taking the local circumstances into consideration, i.e. it has to be designed in detail, but this activity may be executed by other persons than the transferor.

8. TECHNICAL ASSISTANCE

Technical assistance in its most general definition consists of various obligations or a series thereof, which are mainly *facere* (to do) obligations and which may comprise all elements necessary for the realisation of a technical project, starting with consulting. In the course of a technical assistance activity know-how may be transferred, insofar as this is a *dare* (to give) activity. Conversely, the transfer of know-how is often accompanied by technical assistance.

The term 'technical assistance' is also used when we wish to express more than one service with a collective expression (e.g. the seller hands over to the buyer such and such documentation, delivers this and that, and in addition renders technical assistance as specified in the relevant agreement). This expression is to be found in major foreign languages too, e.g. in German: *technische Hilfe,* in French: *assistance technique.*

From the above it is clear that technical assistance is mainly a service. It is also obvious that if the nature of the technical assistance service is not specified in detail, i.e. without ambiguity, this expression remains an unpractical generality.

Since technical assistance is, like engineering, an important tool for bringing technical ideas into practice, we hereunder discuss it in some detail.

Technical assistance may be divided into the following groups:

1. One group is comprised of activities, in the course of which the party rendering the technical assistance puts *experts* at the disposal of the other party, with differing liabilities, in order to perform specified tasks. This activity – if it is a technical one – is mostly, but not exclusively, connected with the erection or putting into operation (starting up) of a machine, production line, plant section, or a whole plant.

It may be:

- the erection,
- technical control during the erection or putting into operation,
- the technical direction of the erection works,
- supervision of a third party performing the engineering,
- the technical direction of the starting of operations, or
- assistance during the operation.

Belonging to this group, but a bit different are the putting of experts at the client's disposal in order to establish an organisation (e.g. an investment, a sale, etc.) and the partial performing of this activity, and the putting of experts at the client's disposal for the performance of a targeted task (investment, maintenance) or for its organisation.

2. Another group compromises those activities which are meant to lead to the *partner who requests technical assistance achieving mastery* over the machine, plant, etc. This concerns activities such as training, consultation, etc.

3. There are some special services, rendered by the party giving the technical assistance, *which can not be performed at all or in a proper way by the* *recipient lacking the necessary expertise.* Examples are certain investigations, e.g.:

- the examination of the usability of raw materials,
- supervision of the production,
- quality control of the product.

4. *General assessment* services are of a different character, e.g.:

- analytical studies (looking for causes),
- surveys,
- investigations of economic efficiency.

5. Finally, various special services not listed above may be grouped. Examples are the lending of certain tools or technical means (temporary loans), and other temporary services.

The above classification tries to outline and order the possible types of technical assistance into various groups. In practice, however, a certain technical assistance type may be accompanied by contributions from several groups (e.g. the supervision of erection and putting into operation, training and certain quality control services).

Most often, though, the services mentioned above do not stand alone, but are coupled to other transactions (e.g. the delivery of equipment together with the supervision of the erection and putting into operation, and know-how transfer with training). The main deal may nonetheless concern technical assistance, e.g. a study or an examination. It may even happen that such technical assistance can later lead to a more ambitious relationship, e.g. a feasibility study is made first, based upon which an engineering agreement is concluded, followed by a transfer of know-how and the delivery of machinery.

Technical assistance can hence be an aim *per se,* but it may also be a tool in order to reach other deals; it can also be a service supplementing other contracts.

Various liabilities are connected to technical assistance services, depending upon whether the legal relation is a simple commissionary one or if it forms part of an entrepreneurial legal relation, and further depending on how the scope of liability is limited in the agreement.

9. NON-TECHNICAL ASSISTANCE

If one looks at the above-mentioned technical assistance categories from another angle, one may remark that nearly all methods of giving such assistance are also encountered when non-technical know-how is transferred. Only the content of the assistance is different: the putting at disposal of experts, training, consultations, appraisals, etc. are just as present, but the experts and the expertise will not be technical ones. These services all promote and are often indispensable for the transfer of non-technical know-how. Thus, everything said above concerning technical assistance is also valid for assistance coupled to the transfer of non-technical know-how.

10. TRADE MARKS

According to § 1 of the Hungarian Trade Mark Act[7] 'any such mark may get a trade mark protection, which is suitable for distinguishing the goods and services (hereinafter: goods) ... from other goods.' A trade mark may be a word, a word combination, a picture, a composition of colours, a flat or spacial configuration, a sound- or light signal, or the combined use of any of the above.

A trade mark is hence a *distinctive mark,* but at the same time it characterises a company and tries to emphasise the connection between the company and the goods. Its distinctive character originates out of this fact. The mark has no distinctive character, if it is used generally for the designation of a certain ware, and if it shows only the ware quality, use, value, place of origin or date of manufacture.

Many marks cannot be protected, such as marks so similar to other marks that they may be confused with them, or if the trade mark is a so-called notorious (i.e. widely known) one. Marks which bear exclusively names, abbreviations, flags, coats of arms, emblems, or official control or quality marks, etc. of a State, of an authority or of an international organisation, may also not be protected.

In order to establish the limits of the scope of the trade mark protection, goods and services are classified into 34 classes of goods and 8 classes of services. One may request registration in one or more classes of goods or services. A limitation of registration is necessary to prevent the entitled person from registering a trade mark for all similar types of goods. It is justified in the case of goods which totally differ from each other to allow the use of the same or of a similar trade mark and also its registration.

The registration of a trade mark has to be asked for. Fees have to be paid for the registration and for the renewal of a registration. The trade mark application is researched *ex officio* and if allowable, it is registered, which means that the proprietor of a trade mark has the sole right to use the trade mark for the (class of) goods shown in the register and to grant a licence for its use to others. The period of protection is 10 years, which is indefinitely renewable in ten year steps.

7. Hungarian Trade Mark Act IX of 1969; most other countries have similar rules for trade marks.

The lack of use makes the trade mark vulnerable: after five years the cancellation of the protection may be requested in Hungary (paragraph (1) of § 35 of the Trade Mark Act). In some countries (e.g. Hungary, Russia, China, etc.) the use of a trade mark only in the form of an advertisement or as a business paper heading is also accepted as its full use. In most countries, however, only its use on the goods themselves or on the packaging is accepted as full use. There are even countries which go further: a condition for the renewal of a trade mark is that the owner should certify its use prior to the renewal (e.g. the USA and Japan).

Under the Madrid Convention one may obtain international protection through an application in one of the 31 European and non-European member countries (Hungary is a member of this convention). For registration in other countries a broader international convention exists, the Paris Convention, which secures the priority date of the first application in 100 countries. It is to be remarked however that the priority date is of less importance in the case of trade marks than in case of patents, but the reciprocity is important. It has to be emphasised again, that in the case of trade marks their *use* is extremely important in most countries of the world.

One may grant a licence for trade mark use just as well as for patent use. The illicit user is an infringer, who may be liable to a prohibition of use and the payment of damages as well as to penal consequences.

It has to be remarked that in the course of a transfer of technology, especially for consumer goods, the transfer of a trade mark (a trade mark licence) can be just as important as the transfer of the technology itself.

A trade name, if used for the marking of goods, is similar in function to a trade mark. Its transfer however is more complicated than that of a trade mark, since it may collide with the principle of exclusivity of a company.

11. INDUSTRIAL DESIGNS

The industrial design,[8] in contrast to the utility model (*see* paragraph 4 of this chapter), relates exclusively to the aesthetic appearance and shape of an article in two or three dimensions. It therefore serves as a protection of the outer form. An application has to be filed to acquire a protection. The period of protection in Hungary is presently 5 years, which may be renewed for a further 5 years (in some countries other time periods are applicable). It grants similar exclusive rights as a patent or a trade mark, hence its use by others may also be allowed (a licence may be given). It is important only when the aesthetic appearance is of special importance.

8. *See* e.g. the Hungarian Law-Decree No 28 of 1978.

12. LICENSING IN GENERAL

The subjects of the transfer of technology, namely patents, know-how, utility models, industrial designs, and trade marks serve a dual purpose: on the one hand they must be used in practice to retain their legal validity, and on the other hand their use may be assigned to others against a proper fee. Such an assignment is called a permit of use or a *licence.*

12.1. Exclusivity

There are two types of licences: exclusive and non-exclusive ones. In the first case the licensee gets full rights to utilise another's (namely the licensors') patent, know-how, etc. In case of non-exclusivity the proprietor (the licensor) may give a licence to others too (i.e. several licensees may exist).

The meaning of exclusivity may be ambiguous, so the parties should clearly define it in the agreement. The question is first, whether the exclusivity is so strong that it excludes the licensor from the use, or such that it does not exclude the licensor from the use but only excludes the granting of further licences.

The other question is that of geographical limitation. The granting of the registered rights (a patent, an utility model, an industrial design) and in case of the trade mark, its registration are bound to the jurisdiction of a country. Should the rights be granted for more than one country, then the granting of a licence is possible for certain or all countries. Sometimes the grantor does not give a licence for its home country, but only for other countries. In this case partly exclusive and partly non-exclusive rights (this may even vary from country to country) can be transferred, as agreed.

The situation is similar for non-registered rights, handed over within a frame of an agreement by the grantor to the grantee for one or more countries. The exclusivity in this case means only that the proprietor (grantor) promises not to grant the same right to others. If the licensing relates to more than one country, the grantor may however fix the circle of countries for which he grants exclusivity.

12.2. Subject matter of the licence

Another method of classification could be for what the rights of use relate. Except when it concerns a trade mark, the licence may allow the production, the bringing to market (the sale), and the bringing to market in third countries (i.e. the export) of goods or services. The licence may also allow the grantee (licensee) to pass the licence on further (to grant sub-licences). The trade-mark licence – due to the fact that its subject is the use of a mark on goods – will obviously extend only to the bringing into circulation, namely the sale and/or export.

Another important point is whether the transfer of the licence is the main and/or only subject of the relevant agreement, or is only one of the contractual collateral obligations of a major contract, such as e.g. a joint venture agreement (*see* Chapter 5). A technology transfer is hardly feasible without the transfer of some of the above-mentioned rights (i.e. some sort of licence).

12.3. Compulsory licences

In connection with the patent we have already mentioned that the patentee has to use his patent. Should he fail to do so, a compulsory licence lawsuit may be initiated against him. The legal consequences of non-use may however be avoided in various ways. No compulsory licence will be granted, if:

- one has made serious preparations for the utilisation of the invention in the country in question,
- a right of use (licence) has been given to others,
- the patentee has a valid excuse for the non-use (e.g. he advertised in the professional press that he was willing to grant a licence – under proper conditions – for his patent, but without timely result.)

Not only lack of use may result in a compulsory licence lawsuit, but also if one's patent cannot be used due to a dependence on the patent of someone else. The granting of a compulsory licence from the latter may then be requested. This dependency cause is more important than the cause of non-use.

There are certain constraints in the case of a compulsory licence. For example, the licence may not be an exclusive one, the compulsory licensee may not grant a sub-licence, and the patent on which the compulsory licence is based is to be used within a certain period of time (one year from grant in Hungary).

2 The Modalities of Transfer

1. FIXED (WRITTEN) TRANSFERS

The transferor – as already shown in the previous chapter – may undertake more than one service within the frame of a single contract. The nature and extent of the services have to be evident from the agreement(s). Besides determining unambiguously the type of service (e.g. a patent licence agreement, a know-how transfer agreement, an engineering activity such as basic engineering and/or detailed engineering), it must also be stipulated in detail what sort of technical and non-technical documents are to be handed over by the transferor, what is to be their extent, and in which language they are to be prepared. In the case of technical documents, the standards to be used have to be stipulated and it must be clearly stated how the delivery of the documents will take place.

1.1. Documentation in general

There are different sorts of documents. Without giving a complete list but to give an idea, the following documents which serve mainly to define the subject of a technology transfer may be mentioned:

- technical descriptions in general,
- production descriptions (so-called technological descriptions),
- technical drawings and designs (plans) in general,
- technical specifications (of a product, machine, etc.),
- data sheets, lists of material,
- prescriptions for material compositions (formulae),
- testing results in general,
- the results of chemical analyses,
- technical calculations,
- manuals of production and technical prescriptions (e.g. manuals which contain a description of a machine and working and maintenance instructions),
- production or product-control descriptions,
- quality control results,
- trouble-shooting instructions,

- optimisation programs,
- technical models (even scale-down models of a plant),
- technical reports and general notes,
- research reports.

The subject of the transfer is usually contained in documentation of several sorts.

The so-called complete documentation may contain a great variety of technical information, part of which may be in blueprint form and the rest in reproduced or printed text form, accompanied by technical drawings, possibly detailed, technical calculations, tables, diagrams, etc. It is impossible to give a complete list of the types of documents which can be involved in a transfer, but it is worth noting that written know-how documents contain more details than patent descriptions and are also more extensive than patent specifications. A patent description under normal circumstances only contains a few pages and its drawings are sketchy. By contrast, documents relating to a transfer are much more voluminous, containing many hundreds or even thousands of pages. The drawings belonging to the transfer documents are usually proper technical drawings, and even when they are not the relevant dimensions are clearly marked. A complete documentation is characteristically furnished with a lot of detailed drawings.

Documents and drawings may be in a 'non-classical' form, such as on a computer disk, making the required information accessible and reproducible at will.

A non-technical transfer, too, may be set down in various documents. Due to the large variety of such documents, we shall not itemise them. We mention just as an example that in respect of a transfer of commercial and financial knowledge or in connection with the operation of a hotel, various non-technical documents may be handed over (*see* paragraph 6 of Chapter 1).

An important question is the *language* in which the documentation is phrased, first of all in respect of the text (a drawing may contain two texts, one in the language of the transferor and the other in the language of the transferee). If the transferor's seat is in a country of a world language (German, English, French or Russian) it is usual for the documentation to be delivered in the language of its home country. If there is any danger of ambiguity in its interpretation, the transfer contract should state clearly the language used for the various kinds of documents. Any translation of the documentation from the stipulated language is made at the cost and risk of the transferee.

In technical documentation it is an important issue, what kind of *standards* have been or are going to be introduced, at least in Europe. Should there be no unique official standard, it has to be agreed to which standard the documents (e.g. technical drawings) should correspond, first of all as to their content, but even as regards their form.

Official standards may be either those of the machine producer's country, the transferor's country, or the transferee's country. In some cases national standards or recommendations generally used in the technical field concerned (e.g. in the petroleum industry ASTM prescriptions), are to be used.

In the course of the transfer the related documents must also deal with environmental and safety regulations which protect both the working personnel and the nearby population. The periodical or continuous check of some devices, the monitoring of emissions and medical examinations come under this heading.

If the transferee does not furnish the safety regulations applicable to the site, the transferor has to furnish them. If these are insufficient for the chosen site, the transferee has to point out the added requirements and make the transferor liable for presenting a complete plan in this respect, upon of course a special surplus fee.

In the transfer contract the form and/or number of copies of the documents is to be set down. A single copy of the document would rarely be sufficient, since there are often more than two parties involved in a transfer, such as, besides the transferor and the transferee, the transferee's designer. In any case a practical solution would be to deliver the documents simultaneously in several copies, as asked for.

The *means* and the *parity* also have to be clarified in connection with the documents to be delivered, similarly as for the delivery of goods.

The following have to be set down in an agreement in connection with the documentation:

a. the terms of delivery,
b. the place of performance,
c. the time of performance,
d. the destination,
e. the transfer of the risks,
f. possible insurance,
g. other stipulations.

Ad a: The *terms of delivery* of the documents may be prescribed in various ways. Depending on the extent of the documentation, one or more terms of delivery may be stipulated. It is usual for the documents to be delivered at different times, since the transferee may have to furnish further data to the transferor later on. The transferor, however, may start the work only after he is in possession of all the main data. The situation may get more complicated if the designing work is split between the transferor and the transferee and/or its subcontractor(s). (We shall discuss such cases later in detail.)

Should the transferor prepare a basic and detailed engineering for the transferee, the preparation and the delivery of documentation takes place in at least two steps. First, the basic engineering is completed and delivered, followed by a discussion – a so-called plan coordination meeting – and only thereafter can the detailed engineering be prepared and delivered.

A delivery in phases of the documentation may also take place in some other cases. For example, when the transfer is connected with a major investment, in order to save time those documents are needed first which pertain to the construction or the reconstruction of the buildings, the remaining documents being delivered later.

The sequence in which the designs have to be prepared is determined by necessity and circumstance. In the case of step-by-step designing it is very important, however, to fix the delivery time for each step in the contract. A number of terms – especially in respect of the detailed engineering – depend on the progress of the transferee's activity, and therefore the dates fixed cannot be regarded as absolute. This may later cause misunderstanding. Nonetheless, the scheduling and its contractual fixing are still necessary and careful scheduling may result in the shortening of the whole designing period, naturally with many compromises but with little misunderstanding.

In connection with the *way* of delivery, experience shows that EMS or airfreight delivery should be used instead of parcel post in the case of an international transfer. The air companies provide bills of lading to the supplier, while parcel post deliveries go through custom houses and are therefore much slower.

It has also to be remarked that due to the almost universal availability of telefax, documents may be sent in this way. They may of course also be delivered to the transferee's address by the transferor personally. In the latter case the delivery must be certified.

In whatever way the delivery takes place, the *place of destination* and the *delivery address* have to be fixed in the contract, as well as how the individual parts of the documents are to be identified.

Ad b: Under the *place of performance* that place is to be understood – in this connection with respect to the delivery of documents – where the transferor has to fulfil its obligation of delivery of documents and where the transferee is obliged – in a contractual sense – to accept the delivery. The delivery of documentation is effected in due time and accurately, if it arrives at the place of performance before or at the time prescribed in the contract. This place is not necessarily identical with the address of the transferee. In case of the delivery of documentation one may stipulate as the place of performance:

- an airport (that of forwarding or that of arrival),
- the handing over to a transport company (an airline company or EMS),
- the handing over to the transferee personally at the transferee's office,
- telefax communication.

Ad c: The *time of performance* is that time when the transferor has fulfilled its delivery obligation. E.g., if the time and place of performance are the airport of Schiphol on 1 February, the transferor has fulfilled its obligation only if the documentation arrives on that date at that airport. It is insufficient to hand over the documents for forwarding to an airline company on 1 February if they do not arrive in time.

Ad d: The *destination* (of transport) indicates, which party shall cover the transport costs up to which point. Should the transport destination be an airport in North America, the transferor is obliged only to cover the transport costs up to that airport. The destination of transport is in most cases identical with the place of performance, but not always.

Ad e: *The transfer of the risks* is regulated in different ways in various countries. According to Swiss substantive law the handing over to the transport company is decisive for the transfer of the risks, independently from any other factor. According to Hungarian law (paragraph (2) of § 279 of the Civil Code) and many other national laws, the risks pass from the transferor to the transferee with the performance. Should the place of performance be different from the delivery address (e.g. if the place of performance for the delivery of documentation is a North American airport, but the transferor undertakes to deliver the documentation to Budapest) the risks pass from the transferor to the transferee at that time when the documentation is handed over to the airline company (or the transport company) already at the North American airport.

In the case of telefaxing this question is much simpler, because the sending and the receiving happens at the same time.

Ad f: Concerning *insurance*, it has to be remarked that it is not necessary to insure the documentation for the period of transport as it would be for the transport of goods, because in case of loss or destruction the substitution can be effected by making copies of the original; hence the insurance is of secondary interest. But if the delivery of documentation is coupled to the delivery of goods, the former may be insured at a little surplus of the insurance premium.

Ad g: Under *other stipulations* one may consider *completeness*. The transferee often does not know whether the documents enumerated in the agreement are sufficient for a successful transfer or whether an element was omitted (probably by mistake). Stipulating that anything missing but necessary for the realisation of the transfer should be delivered by the transferor even if it is not explicitly contained in the agreement, could cope with this situation. Such a stipulation may be called a 'declaration of completeness.'

The possibility has to be reckoned with that something important was omitted in the enumeration of the documents, since the transferor could not have any idea of it, the transferee having mentioned it only after concluding the agreement. For example, if the subject of the transfer is a product which is to be prepared from a certain raw material, and it only appears after the conclusion of the agreement that the transferee does not possess the required raw material in a sufficient quantity and/or quality, the concept of completeness has no meaning and the transfer agreement has to be modified. Upon this modification the countervalue of the transfer may also change, as well as the kind of the documentation to be delivered, its terms of delivery, and many other conditions which are connected with the modification requested by the transferee.

1.2. Drawings

In the course of a transfer of technology a number of drawings may be handed over. Without dealing with these drawings in detail, we mention only that the drawings are mainly:

- general layout drawings,
- flow sheets (e.g. the flow sheet of chemical processes),
- assembly drawings,
- designs of detail, and
- sketches of the appearance (view).

They may also be drawings of a technical branch, such as:

- mining designs,
- mechanical designs,
- electric or automation designs,
- network designs (road, piping, sewage, etc.),
- buildings (surface construction, civil engineering), or
- explanatory drawings.

1.3. Machine manuals

Even in the case of buying a machine, a low-grade technology transfer takes place, because the machine is necessarily accompanied by a machine manual. Besides the technical description of the equipment in question, the machine manual also contains the data for its operation (description of the operation and maintenance), and even the most usual failures and their eventual correction. In the case of a simple machine, there is no need to instruct the buyer on how to operate it, as he finds everything in the machine manual. In the case of more complicated equipment the situation is not so simple. Here a more detailed description which reviews all the main elements of the composite equipment is required. Often certain parts of the equipment are not manufactured in the same workshop, and the final deliverer of the equipment may then choose between synthesising the elements of the machine originating from various places into a single body of knowledge, and – if this seems more practical – retaining the individual product information sheets and arranging them into a single volume. There are cases when the equipment is so complicated that this is impossible. In such cases the mechanical equipment, the electric equipment and the instrumentation each make up one documentation segment and are handed over in this form, assuming that the transferee has at its disposal specialists who can deal with such documentation.

1.4. Models

A part of the transfer may be embodied, in lieu of detailed drawings and descriptions, in a technical model (e.g. in a mock-up of the plant with a model size of 1:100). Such a model may give information on an individual topic (e.g. about the piping design of the plant) which is just as detailed as a description and/or drawings amounting to several hundred pages.

Naturally it is of little use if the model is available to the transferor only, and it is usually made in two copies, one of which is for the transferee. It is also possible that a model of e.g. piping is necessary only at the site and that the transferor delivers it to the transferee without keeping a copy. In any case, to avoid later misunderstanding, it has to be stipulated in the agreement that a model is to be made (e.g. instead of some drawings) and that it shall be delivered as part of the documentation.

The transport of the model is not an easy issue, since its dimensions must remain unchanged. The smallest dislocation may change the dimensions of the model and this may cause difficulties at the site.

1.5. Data furnishing, task sharing in designing

In connection with a transfer the question of *initial data* is very important. In case of a transfer involving an investment, one has to fix all those data in the agreement which serve as the basis of designing. Hereto belong the data relating to the site, the raw materials, the auxiliary materials, the types of energy to be used and other important factors. (It is e.g. impossible to design an alumina plant and to guarantee the proper result, if the quality of the bauxite is not determined in advance).

It is practical to set down the initial data in an appendix to the agreement. The later furnishing of data, which are not available at the time of contract signing, by the transferee may – depending on their importance – be made a condition for the coming into force of the agreement, but if they are not transmitted within a certain time this may have an effect on the terms of delivery of the documentation. For example, the lack of soil examination data may not only hinder the building and civil engineering but also the general layout of the plant and so the planning of the entire investment. The availability of such data in some cases is a precondition for starting the designing work.

It has already been mentioned what is generally understood under basic engineering in the case of a transfer bound to an investment (*see* paragraph 7.1

of Chapter 1). In specific cases, though, more or less information is required (e.g. the general specifications of the equipment are sufficient, but for certain parts thereof more data are required). The situation is more complicated if the transferor supplies engineering which is basic in some respects, but detailed in other respects. This is often the case if the transferor and the transferee share the task of designing. Usually the transferee does not ask the supplier of the technological engineering to carry out the building and civil engineering, but does it himself or orders it from some other engineering company. (In both cases these activities are within the liability of the transferee.) In such cases it is usual for the supplier of the general (technological) engineering to furnish only simple data for the building and civil engineering designing, or to supply the conceptual designs (which is similar to the basic engineering). He will probably also participate in the designing work of the transferee (or the other engineering company) through a consulting engineering activity, e.g. by reviewing the final designs. Even if this latter activity is not detailed in every respect in the agreement, the main activity to be performed must certainly be fixed.

In case of a transfer bound to an investment, task sharing occurs also outside the building and civil engineering activities. The transferee often limits the activity of the supplier of the engineering to within the boundaries of the plant, some external connections however being necessary (e.g. the electric network). In such cases it has to be fixed in the agreement itself, whether e.g. a transformer is to be regarded as being within or outside the boundaries of the plant and who is to do its designing. Even the concept of *boundary* may be ambiguous in some cases. It may mean the line between the strictly basic technology and the auxiliary equipment, which the transferee may order from a third party.

Task sharing may be possible in the designing of some equipment itself. If the transferee wishes to produce a simple piece of equipment himself (e.g. a storage tank), it is practical if the detailed engineering thereof is also made by the transferee; in other words it is not performed by the supplier of the engineering (the transferor), although it might be within the boundary limits. It is practical however to request the basic engineering of such a piece too from the transferor.

Task sharing may take place not only for simple equipment, but for complex equipment too or for certain kinds of equipment (e.g. electric equipment) the designs of which are not to be made by the supplier of the engineering (the transferor). The general idea in such cases is, that the basic engineering should be fully performed by the company preparing the technological design (generally the transferor), while the detailed engineering should be made by the party supplying the relevant equipment. Should the supplier of the engineering also be a main contractor (furnisher of the equipment), he will probably execute the detailed engineering too, at least in respect of those parts which are within the scope of his delivery, while the transferee takes charge of the detailed engineering of the equipment which is to be manufactured by him.

A practical remark: it is usual to stipulate the basic issues in connection with the designing (engineering) in the main text of the agreement and to regulate the details in connection with the documentation to be delivered in one or more appendices. By 'details' are also meant the number of copies to be delivered, the language of the documents, etc.

It has to be remarked finally that in case of task sharing in designing (engineering) only a step-by-step designing is feasible, and the designs have to be coordinated after each step (*see* paragraph 2.3.2 of this chapter).

2. THE NON-FIXED TRANSFER

It is more difficult to describe the presentation forms of the non-fixed transfer. A general feature of this category, it may be said, is that it contains unwritten industrial knowledge, experience, practice, etc. Should the subject of the transfer be know-how and the transferor intends to transfer it by other means than with written documents, the form of the transfer may be, e.g.:

- experience collected during visits to plants,
- knowledge acquired by training,
- knowledge acquired in the course of technical discussions, exchanges of ideas, or pooling of experience.

Due to the fact that the actual handing over of the know-how is in these situations mainly complementary, a small significance is often attributed to this type of transfer. This idea is however wrong. The non-fixed transfer not only occurs as an accessory to the fixed one, but also often occurs by itself and hence it is just as important as a fixed one. When drawing up the contract, one must consider that this type of know-how transfer requires very careful formulation.

2.1. The show-how

No training is included in the concept of show-how. Often a simple visit to an up-to-date plant may yield information which the transferee can put to use in its own plant. Sometimes, too, the transferor is willing to transfer its know-how only in this way.

Know-how agreements have been contracted, where the sole subject of the contract was that the grantor of know-how, the transferor, allowed the transferee against remuneration to visit and/or study for a longer or shorter time its plant, without undertaking any other service. The transferee then has the right to utilise the techniques seen in the plant in its own plants: hence this solution is called 'show-how.'

It has to be emphasised that such a form of the transfer is very comfortable for the transferor, because he does not have to prepare and deliver any technical documents and his liability is practically nil. For a potential transferee this solution is feasible, if he is properly prepared to apply the knowledge acquired in the course of a simple plant visit. It is obvious that this form of transfer is not interesting for a not properly skilled transferee.

2.2. Training

In the course of training the transferor, the grantor of the know-how, trains (teaches) the experts of the transferee in a production process, in proper operation, in trouble-shooting, in laboratory investigations, or in any other subject, in general against a considerable fee. This training may take place in the plant of the transferor or in the headquarters of his company, or at a place chosen by the transferee (in its plant, laboratory, offices or at the site of a plant under construction, etc.).

Although the training has a complementary significance only, it is nevertheless important. Many production processes do not have the required result without an adequate training. Such was the case e.g. in the production of a special fibre, where the workers had to be trained in the proper care of their hands (a special art of manicure) to be able to do their work in the factory properly. Training similarly serves for the acquisition of the necessary practical experience, for the proper operation, or for trouble-shooting.

We have already mentioned that the training may be organised either at the transferor's or at the transferee's premises. The practice is generally that adequately prepared experts (not only engineers, but foremen and skilled workers as well) take part in a training at the transferor's premises for a predetermined period, in the course of which they receive theoretical teaching and practical instructions and even work with the equipment. Thereafter – again for a predetermined period – further training takes place mostly in the transferee's plant under construction or under modification, or at another adequate place. This latter part of the training concerns the acquisition of practical knowledge. It takes place during the period of construction and/or the putting into operation. The transferee's trainees get acquainted with the construction and erection of the equipment in the former case and with the operation in the latter case.

2.3. Technical meetings, discussions

2.3.1. General

Technical discussions, exchanges of views, exchanges of experience (in short: consultations) may also give rise to a serious transfer, mainly between partners on an equal or nearly equal level of development. For beginners this procedure is not of much help.

A transferor with great experience (e.g. the experts of a big concern) may solve technical problems for the experts of a smaller company with less experience, which without such aid could only be solved at greater cost and over a longer time. The transferor usually first visits the plant and the consultation itself only takes place thereafter. The gains of the consultation may be so significant, that it can be worthwhile to pay for it even if such gains are not guaranteed.

The transfer does not generally end with the consultation, since during the plant visit the transferor sees what types of transfer the transferee really needs and may offer them to the transferee. It is possible, therefore, that the consultation will be followed by the delivery of documentation or an engineering activity.

2.3.2. Plan coordination meetings, modifications

A special type of consultation is plan coordination meetings. One needs plan coordinations in practically every case of task sharing of designing, but their frequency will depend on the complexity of the task partition. Broader task sharing in designing generally requires more frequent and regular plan coordination. The frequency of the plan coordination meetings is also influenced by the number of steps in which the designs are submitted. In cases where the designing is performed in many small steps, the frequency of plan coordination meetings may be offset by their shortness; by contrast, a discussion on the whole basic engineering job is more difficult and sometimes longer.

Even if there is no way to stipulate the plan coordination meetings in detail in an agreement, some basic issues in connection with them have to be clarified in any case. These are:

- *The place of the meeting.* This may be at transferor's or transferee's premises as practicable, but may also be held elsewhere. It has to be agreed, however, who shall decide, and what should be the guiding principle in respect of the place and the time of the meeting.
- *The costs.* It is normal that each party covers its own costs, but it is also possible for the transferor to limit them in man-months and/or frequency. Should the transferee request meetings over and above said limit, transferor may ask for an extra payment therefor, provided the meeting does not take place in its own premises.
- *The subject.* In case of a series of scheduled plan coordination meetings, which are fixed beforehand, it is practical to fix the subject of each meeting too. For some transfers it seems to be advantageous if only a list of ques-

tions to be coordinated is drawn up, the time and place of the meetings being agreed later.

It is further necessary to stipulate in the agreement that the results of the plan coordination meetings are to be laid down in *minutes.* These minutes should not be in contradiction with the conditions of the agreement. If, however, the meeting suggests that a change of the original contract terms is warranted, an amendment to the agreement may be drafted and signed. During plan coordination meetings it is namely possible to modify the designing and other delivery terms, in case of deliveries of complete plants the scope and specifications of the equipment to be delivered may be changed, and/or there may be financial effects. In all these cases an amendment of the original agreement is required. In practice, modifications which influence the interim delivery terms of the engineering only and which are otherwise of minor importance (and thus not conducive to a formal amendment of the agreement), are laid down by parties in the minutes of the meeting. Thereby the minutes – without the parties consciously realising it – can imply a *de facto* amendment of the agreement.

Although minor modifications do not really need to be drafted into an amendment of the agreement, considerable practice is needed to judge whether a question which is regarded as unimportant at the time shall not later grow into a more important one.

Modifications are unavoidable in the course of designing work. They may be due to different causes which are relevant in order to establish whether the modification has financial consequences.

Should the transferee reserve its rights to be consulted about the designs received and to approve them, or if the transferor asks for same, the unavoidable consequence of this procedure will be the modification of the already delivered designs. This will surely have an effect on the entire designing period (delivery terms). If the final projects are conditional on the approval of the preliminary designs, the delivery terms of the final designs will also be dependent thereon.

The rational solution is for the parties to lay down amendments to the basic engineering agreement only in the relevant minutes, which will thereby become a part of that agreement. Thus there will be no need to make an entirely or partly new version of the basic engineering agreement, as everything set down in the minutes will be taken into account in the preparation of the final projects. (This is however not valid where the basic engineering has to be redone for conceptual reasons.) Such amendments are the natural concomitants of the engineering activity, hence their possibility is to be taken into consideration when drafting an agreement.

A quite different issue is when the transferor (the deliverer of the engineering) amends projects which have already been delivered. Some minor amendments are unavoidable, but these generally do not influence the activity of the transferee (e.g. if an already delivered detailed engineering design for equipment is modified but this falls only within the scope of the transferor). It occurs however quite often that an amendment influences the transferee's activity (e.g. architectural designs have to be modified, he has to amend the orders placed by him with a cost surplus as a consequence, the foundations or buildings have to be rebuilt, etc.). The financial consequences of such modifications should be borne by the supplier of the engineering (the transferor), but this has to be stipulated in the agreement.

Due to the fact, that it is not only the right but also the obligation of the supplier of the engineering to incorporate technical modifications, which became available between the signing of the agreement and the supplying of the final engineering plans, into the engineering, some transferors attempt to declare other types of posterior modifications as falling into this category. The truth is that some posterior amendments of the engineering may in effect be manipulations to hide a delay in the delivery of the engineering (i.e. the transferor fulfils in formal respects its delivery obligation by supplying non-suitable designs, and delivers the final, technically correct designs later, claiming them to be amended ones).

Another reason for amendment of the designing may be connected with a declaration of completeness of e.g. an entire plant delivery agreement, namely when one of the parties forgot to explicitly mention something to be delivered in the documents and/or in the basic engineering itself.

Modifications are often quite important, so the issue is sometimes raised in the courts too. The Hungarian Supreme Court (No 42/1988) has e.g. ruled that the party placing an order has to bear the extra costs even in the case of a lump-sum offer, if the original plans were amended by him. (The amendments were asked for after announcing the tender results). The above decision is not only characteristic of the Hungarian system; the legal consequences of amendments appearing in the course of or after designing are substantially similar all over the world.

It is hard to find a text for an agreement which would serve the interests of both parties equally well, but as a general rule it may be fixed in the agreement that should an amendment lead to extra work by any party, the extra costs arising therefrom are to be borne by the amender. This seems to be a pragmatic approach and it protects the ordering party against any *mala fide* amendments by the supplier.

Amendments may also be agreed by the parties, in which case it has to be simultaneously decided who will bear the costs. The party placing the order (the transferee) may also ask for an amendment, for example if his initial data have changed (e.g. the planned raw material is not available). In such cases the engineering work already performed may have to be thrown out in full and started again, leading to extra costs and the amendment of the delivery terms.

2.4. Expert know-how

The consulting activity is gaining in significance nowadays. This activity may be limited to a single person (the expert), but may also be the activity of a whole team (an office). In the former case the expert is transferring his own personal knowledge, in the latter case an office or a team of experts undertakes to solve a technical-economic task, to point out its general lines, or to direct its execution.

Should an office be committed with the solution of a technical task, it will most probably work it out in a fixed (written) form (*see* paragraph 2.1 of this chapter). This may be followed by a consultation (*see* paragraph 2.3 of this chapter).

The case is however different if an expert in a given technical branch is employed for a shorter or longer period. The aim of such employment is namely the acquisition of know-how: the employed expert is expected to solve the task in question by using his personal know-how or by transferring it to the employer.

The situation is similar if an independent scientific expert is temporarily employed. This may however lead to legal problems if the expert is not 'independent' but was previously the employee of a competitor. Such an expert may in effect hand over not only his own personal know-how, but the know-how of the competitor, probably much cheaper than it could have been bought from the competitor – if the latter would be willing to transfer it in the first place. That action could amount to unfair competition with all its legal consequences.

2.5. Contractual conditions for services

The contracts (agreements) in connection with services, as well as those parts of other agreements which relate to services, can be best characterised by some interrogative pronouns such as: what, where, how, when, how many, and for how much.

1. The *what* question relates to the subject of the service. As mentioned already, the collective concept of technical assistance or technical services is suitable for grouping together various services at the drafting of a contract. The concept, however, has little meaning if the agreement does not detail the services one by one (although certain services may also be grouped). Each service has to be stipulated at the right place of the agreement (e.g. in an agreement relating to a complete plant, the supervision of the construction, the starting of operation, the training and other services may be assigned to different chapters.) Should the service form only a collateral obligation, it is often sufficient to fix its subject in the main text of the agreement and to stipulate the details in one or more appendices to the agreement.

2. The *where* question is sometimes self-explanatory (e.g. the supervision of the construction may be done only at the site of the plant). The situation is less clear in the case of training, which may either take place at the site of the plant, or in the factories of the supplier, or at a place designated by the supplier (e.g. if the supplier is a main contractor who is not the owner of the technology, then the training will most probably take place on the premises of the subcontractor furnishing the technology).

In certain cases the question *where* cannot be answered exactly at the conclusion of the agreement (e.g. in the case of training: in a factory to be selected later by the supplier). Here it is advisable to fix in the agreement the exact date on which the decision is to be made.

3. The answer to the question *how* may be quite complex. In cases where the aim of the service is not the providing of experts, but the performing of a certain task (e.g. an investigation), the task to be done has to be clearly defined. For example, an expert team may be contracted to collect some data in the plant of the mandant as a first step, then to process them, and finally to write a report and submit it to the mandant.

Should the subject of the service be the sending of experts, their working time, overtime, and holidays should be fixed – amongst others – in the agreement. It is usual to adapt oneself to the law and practice of the site of the service, but due to the fact that these can be different in the various countries, it is practical to stipulate every element of interest in the agreement.

4. It is not always possible to answer to the question *when* by giving a factual date. Supervision of the construction of an installation can be commenced only when the construction starts, but it is sometimes illogical to fix the moment beforehand.

Such uncertainties can be cumulative in all subsequent technological phases. The parties to the contract often agree only on the probable time and duration, stipulating at the same time that the precise dates for each phase shall be determined later. It is however advisable to set a time limit to the word 'later.'

It happens often that the timetable is fixed in such a way, that x days after a certain event another is to take place (e.g. the parties shall have a meeting 80 days after the commencement of the deliveries in order to decide on scheduling the supervision of the construction). Sometimes maximum periods are determined (e.g. the investigations shall be ready within y months of receiving the samples).

For some services, such as studies and investigations, the timetable or the time limit may easily be fixed in engineering or delivery of goods contracts.

5. The question *how many* is meaningful when sending or receiving experts. The overall number of experts as well as their number in different phases of the work is to be fixed in the agreement. In case of a large-volume service the number of man-months is the only practical one to appear in the agreement.

When deciding upon the total extent of expertise services it is also to be agreed, which one of the parties has the right to call in experts, or whether the issue requires unanimity. This issue can be a source of controversy, since the party receiving the service will want for financial reasons to minimise the number of experts to be sent to him, and at the same time will require from the experts a large scope of activity. This runs against the interests of the sending party.

In this connection it is to be emphasised that the depth of the specialised knowledge and the number of the fields of expertise are inversely proportional. The experts' fees sometimes depend on the number of fields mastered and the extent in which they are internationally acknowledged experts.

When the transferor is the main contractor, he generally works with a number of sub-contractors, but these seldom permit experts of another firm to direct or supervise their activities. Experience shows that this issue requires a fine sense of compromise.

6. The question *for how much,* i.e. the price of the service, is quite a complex one and the answer is influenced by various conditions. There are some services, the countervalue of which may be determined as a lump sum (e.g. a study or an investigation). Very often, however, the sending (eventually the reception) of experts is the activity and in this case lump-sum agreements are more complicated and hence to be avoided where possible. A study or an investigation may be combined with the reception of experts, increasing its price.

The countervalue of the sending or reception of experts may be easily calculated on the basis of the duration of the delegation. This may be an hourly wage, a daily wage, a weekly wage or a monthly salary. The various experts of different kinds and levels belong to different price categories.

It is also to be fixed in the agreement how many hours a working day contains, whether an expert's fee is due for holidays, whether overtime is to be paid and at what tariff. It is also an issue how many days of travel are to be recognised as working days.

A lot of details must therefore be clarified when contracting in order to avoid unnecessary disagreements. It is however possible to set a lump sum (average) for the projected man-months.

Incidental expenses may also be added to the countervalue thus calculated. These have to be stipulated in detail even in a comparatively small agreement. Such incidental costs may be the daily allowance to be paid to the experts in local money (differing from the currency of the agreement), if any. It is to be remarked in this respect that countries with a so-called non-convertible or soft currency like to pay such daily allowances to the experts, thus reducing the hard currency fee to be paid to the sender. Further incidental expenses are travel and accommodation costs. In the case of travel expenses, it has to be fixed which class the expert will take and whether he is given the air tickets or whether he buys them himself against later reimbursement. To travel expenses also belong such incidentals as the cost of travel to an airport, should the expert live at a place without one, and/or the cost of travel to the site if this is far from an airport. Such costs are railway or car costs. It is of course possible, that the sender of the experts includes travel expenses in the agreed lump sum and does not bill them separately.

Another type of travel costs are those emerging at the site, the so-called local transport costs. The receiving party often settles this matter by putting cars at the disposal of the expert(s).

When costs of accommodation are to be borne by the receiving party, he may either provide for accommodation himself or pay a sufficient sum to the expert(s) in local money. The advantage of the latter solution is that the expert cannot come up with exaggerated demands in respect of his accommodation. However, the providing of accommodation *in natura* is the only possibility if there are no hotels or other freely rented accommodations at the site.

Incidental expenses could finally be overweight fees, visa fees and other costs which are closely connected with the delegation.

The agreement has to state clearly what countervalues and incidental expenses are to be paid by the party placing the order, in what currency, how and when, what parts are to be provided *in natura*, and against which proofs. It is usual to pay those amounts which are due in the currency of the agreement on the producing of an invoice and documents (e.g. extracts of the expert's diary) certifying the performance of the service in question and, in case of out-of-pocket expenses, against the handing over of the relevant documents within 30 days of their presentation. In the case of longer works it is often agreed that partial invoices can be presented on a monthly or quarterly basis.

The local money – if this is part of the agreement – is usually paid weekly or fortnightly directly to the experts.

It is also possible that the party placing the order does not pay after completion, but during the work too, in pre-determined instalments. This is e.g. possible if the agreement relating to the services is a lump-sum agreement.

7. The *liability* of the entrepreneur or the commissioner for the various services is also to be stipulated in the agreement. It is obvious that in the case of a study this liability is far from being a general one and means only that the commissioner shall execute it with due care and expertise. When the agreement concerns the leading of an operation, generally with the presence of experts, liability emerges *in merito* and it is usually agreed to what extent the expert shall be liable for the correctness of his instructions and for omitting to give instructions, where such omission leads to damages. At the same time it has also to be stipulated if and to what extent the party placing the order has to follow the expert's instructions.

It is also possible that the transferor has a different liability per subject within the same agreement, and that the receiving party has to follow the different instructions to a different extent. E.g. the experts of the sending party may be required to give no more than advice during the architectural and civil engineering work, but to give instructions during the erection of the technological equipment; hence their liability as well as the obligations of the receiving party vary in the two different phases.

8. Should experts be delegated, it is necessary to dispose of some further important questions in the agreement, either in the main text or in the appendices.

First, it is practical to prescribe that the experts shall keep a *diary* at the site. Every important event in connection with the activity should be entered in it, including the performed working time. The entries, when signed by both parties, are conclusive evidence in case of disaccord (if the agreement stipulates so). The main instructions, remarks and questions of the experts should be entered into the diary, whilst the recipient enters his remarks, requests and notes on the activity in general. At least two copies have to be prepared, one for each of the parties.

Second, if several experts of the transferor are active at the site, a *local leader* should be designated. The local leader of the experts must be in contact with the representative of the transferee. Similarly, transferee has to entrust one of his employees at the site with keeping contact with the local leader of the experts. Such contact will be partly in words and partly through the already mentioned diary. In case of large site areas, mutually agreed division leaders may also keep the operative work contact, but the keeping of more than one diary is not practical. When, however, the activity is being performed at several geographically distant sites simultaneously, a leader who should also keep the diary should be assigned by each party at each site. In this case it is advisable to nominate an overall leader on both sides too.

Third, in the case of large-scale or of long duration services, such issues are also to be regulated as e.g.:

- whether the expert may take his family with him and in case of which duration of the service (6 months is usually the minimum),
- whether, when, how often and for how long he is entitled to a vacation, as well as whether his leave is paid and his travel costs are met (this is usual only in case of a service surpassing 1 year),
- whether the transferee has the right to ask for the substitution of an expert by another one, and for what reasons,
- whether the sender has the right to substitute his expert by another one (naturally at his own expense),
- what is to be done in case of illness, whether the expert gets paid during illness and for how long, who pays for medical care and hospitalisation (the best procedure is a health insurance policy),
- in case of language differences, who has to care for the interpreter and at whose cost,
- what offices or other premises the receiving party will put at the disposal of the experts,

- whether the experts may use the telephone, telex, or fax at transferee's expense,
- etc.

Summing up: The agreements or parts of contracts relating to services are quite diverse. Besides issues of principle, they should also contain many practical details, even in the case of agreements of a relatively small value.

It is evident, however, that a number of details cannot be known at the time of contracting for lack of the necessary knowledge and information. The right procedure is to fix in the agreement the range of questions to be agreed later. 'Later' in this connection means either a certain date or the reaching of certain conditions.

3. TRANSFER VIA DELIVERY OF GOODS

It happens often that the transferor, i.e. the licensor or the grantor of the know-how, supports the realisation of the transfer also with the delivery of some goods. We are not speaking here about the delivery of machinery (this will be treated under paragraph 3.1 of this chapter), but about cases where the transferor provides certain *raw materials* or *goods necessary for production* over a longer period, according to the stipulations of the agreement. Such is the case if the transferor delivers e.g. the catalyst necessary for production, or if in the course of the transfer of the production process for a plastic the transferor agrees to deliver some raw or intermediate materials to the transferee. Other examples are to be found in the field of machine or electrical devices production, because it often happens that the transferor reserves his right to the delivery of certain parts. This situation of *delivery of goods accompanying or supplementing* the transfer is connected with certain licences or the transfer of know-how.

In some cases such deals are made between the transferor and the transferee out of practical considerations. The transferee may not wish or may be unable to start with the production of some raw material, a catalyst, a component, etc. since this would be uneconomic in the quantities required, or because his technical skill is insufficient. In such cases he will request the delivery from the transferor, with the eventual risk of being exposed to an exorbitant price by the transferor. Although that risk may be mitigated to a certain extent by market research, it may still happen that the transferee has no choice (e.g. the raw material or component is not manufactured by others, or is patented and transferor does not grant a licence for it). It is not surprising that in such cases the transferor tries to make use of his monopoly to the full.

Another situation is when the transferor supplies his know-how, sometimes free of charge or against nominal payment, in order to secure a long-term contract for the delivery of goods. This is however not a typical case.

In connection with the coupling of the delivery of goods to a transfer it has to be remarked that this may constitute a restriction of the freedom of contracting and be regarded as unfair competition in many countries (*see* paragraph 4.3 of Chapter 3).

It is superfluous to state that the delivery of goods within the sphere of a transfer (a licence or know-how agreement) does not exclude the rendering of an engineering or technical assistance service in addition.

Some contracts provide for paying the countervalue of the licence or know-how, or a part thereof, with the products to be made under the transfer agreement. In such cases the delivery of goods does not accompany the transfer, but is a form of *payment* for the transfer. This solution has consequences for the transferee in connection with the technical guarantees (these will be discussed in paragraph 7.1 of Chapter 4).

Another type of transfer connected to goods is know-how necessary for the *continuative processing of the goods.* In the case of some materials the seller of the goods in question may be willing to transfer a continuative processing know-how to prevent the goods in question being bought from the competition. This is similar to the case of sellers of certain machinery, who are willing to teach the buyer a very broad application of the machinery bought by the latter.

A continuative processing know-how has to be transferred by the grantor in any case, if he produces and sells a material which differs from the usual and which can hardly be processed without his instructions. (Such cases occur quite often with some raw materials and intermediate products in the plastics industry, as well as with dyestuffs for textiles). Nowadays the materials which can be processed only with the help of the seller's continuative processing know-how comprise a growing segment of the market.

There are cases when the continuative processing know-how of a product sold on the market, or at least some steps of it, is reserved by the proprietor of the know-how or is transferred only in a very limited circle. A typical example of this is colour photography. The technology for the developing of the exposed photo material (film) and its processing into a colour film is often kept secret by the producers of photographic materials, to be used only in their central or branch laboratories, and the know-how is not for sale.

3.1. Delivery of machinery

The simplest transfer of technology is the sale of machinery, since the buyer inevitably learns something about the technology too. Here we discuss the case of a *packaged transfer* where the delivery of machinery is the main issue within the transfer of technology. This does not exclude an engineering activity from taking place within the context of a transfer of know-how or any other service.

A packaged transfer agreement does not relate only to the transfer of technology, but also means that the technology to be transferred forms a part of the equipment delivery, being an organic accessory to it. Thus the transfer of know-how is bound to the equipment purchased.

If an important machine such as e.g. one for the production of plastics is delivered to the transferee, its erection, its starting up, the achieving of a continuous run, the maintaining of the required quality of the product, the fine-tuning of the production, etc., is a labour-intensive procedure, often lasting for months even for experts in the field. These machines may be adjusted in many ways (the distance of the rolls, the speed of pulling, the temperature of processing, etc.) and the machine manual will give advice only for so-called typical cases. Adaptation to the local conditions and to the available raw material, in order to stabilise the production and to operate the machine to its full potential, needs special knowledge and experience. The transfer of this experience is a real transfer of technology.

With the purchase of a production line, a packaged technology transfer is always realised. The machines have to be designed for the site, they need an appropriate building with public utilities, as specified, the transferee has to be trained in its working, etc. The more sophisticated the machines – not to speak of complete plants – the more knowledge and experience are necessary for putting them into operation and securing the best working conditions. It can be said that this transfer of know-how is the real transfer of technology.

3.2. Delivery of complete plants

When delivering complete plants the transferor (or entrepreneur) undertakes the following:

- to perform the technological designing;
- to accept liability for the technological parameters and completeness of the plant to be established;
- to produce (or have produced by sub-contractors) all the necessary equipment, machinery, devices, instruments, etc. (hereinafter, in short, equipment) except those which are to be provided by transferee according to an a itemised list;
- to deliver the equipment before the date fixed in the agreement;
- to direct the erection works carried out by the workers of the transferee;
- to furnish the initial data necessary for the architectural and civil building works to be executed by transferee;
- to direct all activities between the completion of the erection of the plant and its starting up, and to be liable for the operation of the plant during a specified period from the completion of the erection (i.e. to be liable for presenting proof that the capacity and technical parameters have been achieved as undertaken);
- to warrant the quality of the equipment delivered by him and to be liable for its hidden faults up to a certain time limit (normally one year from the

commissioning of the plant), but usually limited to 24 months (sometimes 28–30 months) from the last delivery, whichever is earlier.

The delivery of a complete plant is often called a 'semi-turn-key job.' It is the usual deal with developing countries. As mentioned above, the building and erection works may be carried out by the transferee, hence the transferor (entrepreneur) has no obligation to ensure the functioning (commissioning) of the plant by a fixed date. He is, however, obliged to start up the plant within a certain period after the completion of the construction works and to demonstrate the guaranteed performances.

It may also be agreed that the construction works are to be made by the transferor. In this case he is liable naturally for the duration and quality of the construction, provided that the transferee has executed all the prior building works in a proper way before its commencement. Such was the case for deliveries of complete plants to the former East Germany, because of a lack of personnel for the construction works. Even in the case of such a 'complete' transfer, it is however usual for the transferee to furnish some equipment, i.e. he shall take part in the delivery. The motivation is to utilise the production capacity of the home country and also to save some 'hard currency.' The 'completeness' on the side of transferor is then more or less a 'virtual' one, but not without liability. The equipment to be provided by the transferee being based on the drawings and prescriptions of the transferor (except in the case of primitive equipment), the transferor has to undertake technological guarantees also in this case and to accept the obligation of 'completeness' on the condition of the itemised and true fulfilment of transferee's delivery obligations.

Such type of transfer is usual in countries where there is a relatively developed building industry (including architectural and civil engineering firms) and a well-organised and well-equipped installation capacity with the necessary staff of experts and workers, but which do not belong to the forefront of machine technology.

In all such cases the agreement has to be drafted very cautiously and precisely. It is not easy to delimit the mutual liabilities of the parties, hence the negotiating group on the transferee's side must consist of well-trained experts who know all about the properties of the plant to be established and about the adjoined investments. If the negotiating group is incompetent and comes up with modifications in the course of the actual work, the transferor will either refuse such changes or demand payment of a corresponding extra charge.

The price (countervalue) is mostly a lump-sum price for the engineering works and the delivery of equipment. To this lump sum may be added certain budgeted amounts for various purposes (e.g. spare parts, direction of construction, etc.) to be fixed in the contract. To cope with inflation a lump-sum price is often fixed with an inflationary hedge (indexing), determined first of all by the increase of wages, salaries and raw material prices.

The transferor (entrepreneur) who delivers a complete plant nevertheless has to reckon with a number of risks, which he will surely try to incorporate into his price. The price is determined by many elements, such as:

- the price of the engineering,
- the countervalue of the know-how,
- a possible royalty to be paid,
- the countervalue of the delivery of equipment,
- a main contractor's fee,
- the risks relating to the completeness,
- an inflationary risk,
- a risk due to the punitive sanctions in the agreement (e.g. penalties for non-fulfilment of the technological parameters, liability for direct damages, the risk of an eventual technological fiasco),
- extra expenditures because of prescriptions of labour safety, fire protection, and similar measures at transferee's site,
- a risk connected with the period of warranty,
- a possible loss originating from rendering credit to the transferee at fixed rates of interest which are lower than those prevalent at a later time.

Such contracting is not very advantageous, as it contains two eventual sources of tension between the parties: the fixed price and the many risk factors. There is a possibility of substituting the fixed price by a gliding one, causing some uncertainty to the investor (the transferee). A further possibility is that the agreement contains a fixed price only for the engineering and the know-how, with the entrepreneur-transferor sending the invoices of his sub-contractors to the transferee for payment. This latter solution is a so-called 'open cost' one.

More usual is a so-called 'cost plus fee' solution, the essence of which is that the transferee places an order for the engineering works and the accompanying know-how transfer with a company found suitable for this job (in most cases an engineering company) while purchasing the machinery equipment, etc. according to its discretion, maybe taking advantage of the technical advice of the engineering company providing the know-how.

A variation on this type of contract is an engineering agreement supplemented with an order for the main equipment from the transferor. In such cases the transferee buys only the auxiliary equipment from others, or produces it himself.

In a cost plus fee agreement the transferee orders only the process and the technical designs up to a certain degree. He may also procure some equipment from transferor, but essentially *he is his own main contractor,* because the harmonisation of the details, the adaptation to home country prescriptions, the elaboration of details and the procurement of the machinery is his task.

An advantage of this type of transfer is that it seems to be cheaper, and the experience of the transferee may be inserted in any stage of the project. It is obvious that this form should be used where the transferee has sufficient experience in the given area but lacks an up-to-date technology.

This way of contracting does not necessitate the incorporation of a detailed specification of the plant in the agreement, since the plans develop over time as the transferor hands over his designs. Nevertheless the transferee has to monitor the whole procedure of technical designing to ensure that the transferor performing the engineering does not deviate from the agreed requirements and does not design a plant which shall finally be similar to a 'golden cage.'

3.3. Turn-key plants

The essence of a turn-key transfer is that the transferee puts a site – possibly an enclosed area – on which a plant is to be established at transferor's disposal and the latter undertakes to construct an establishment there. At the same time he guarantees the technical parameters of the plant and the date of commissioning in the agreement.

In this scheme the transferor-entrepreneur undertakes the following:

- the full engineering,
- the execution of the building works (architectural and civil engineering),
- delivery of all equipment, machinery, devices, materials, instruments, etc. necessary for the plant,
- erection of the above,
- starting up and adjusting the equipment,
- putting the establishment into operation, giving proof that the parameters guaranteed are reached and handing over the plant in due time.

The transferor's liability is the greatest for turn-key contracts. Such a deal practically excludes the transferee from the execution of the works, denies to him the related experience, and forces him to buy from the transferor all necessary materials and semi-products which might also be available in the country of the transferee or might be procured from third parties under more favourable conditions.

The turn-key way of buying is customary in developing countries, where the transferor does not wish to be informed what is available at the site and what is not. This is the most packaged form of transfer of technology. It has many drawbacks for the transferee:

- the developing countries have to pay in hard currency for the architectural and civil building works,
- the price is high, and often a dictated one,
- the accommodation, including board and high quality service, of a large number of foreigners during a relatively long period may be required at an

up-to-date accommodation facility (sometimes there are several thousand people present at the site during the peak building period),
- although the transferor-entrepreneur guarantees in the agreement the date of handing over of the plant, he may impose as a condition that some services of the transferee are to be performed in due time, the non- or late performance of which may exempt transferor from this obligation.

Against the enumerated drawbacks, the main *advantage* of this type of transfer is that it allows – at least in principle – *quick realisation* and relieves the transferee from any anxiety.

4. TRANSFER OF SYSTEMS

While the transfer of complete plants is characteristic for the transfer of complex know-how and usually concerns industrial plants, the transfer of systems is characteristic to agriculture and the so-called tertiary (service) sector.

It is not easy to deal with the transfer of a whole system in a single contract. Most often a blanket agreement is concluded, enumerating the future transfers and containing some general regulations for the transfer. In respect of the details, i.e. for the individual objects of transfer, separate agreements are concluded (it would e.g. be problematic to stipulate in an agreement concerning a seed-treating plant conditions for the delivery of sowing seeds). This type of transfer contains both technical and non-technical elements.

It is to be remarked that while transfers in the field of crops and livestock are directed mainly towards the developing countries, the transfer of services is more universal.

5. TRANSFER OF NON-TECHNICAL KNOW-HOW

The technical types of transfers enumerated above (*see* paragraph 1.2 of this chapter) are in principle applicable to non-technical know-how as well. The putting at disposal of experts, training, consultations, investigations, etc., are all services which also promote the transfer of non-technical know-how and are sometimes even indispensable for the proper acceptance of the transfer.

Due to this everything which has been mentioned in connection with technical documentation and assistance, also holds for the transfer of non-technical know-how, although in the case of non-technical know-how the transfer of the knowledge and experience of experts is relatively more important.

3 Some Important Legal Issues[1]

1. POSSIBLE MEANS FOR THE PROTECTION OF TECHNOLOGY

Both legal and natural persons may own intellectual property. An intellectual creation stands under the protection of the law, as discussed in Chapter 1.

1.1. Legal protections

Here will be discussed what legal protection possibilities are available to creations which fall under the concept of technology.

a. Patents
The concept and essence of the patent was clarified in Chapter 1. Here we emphasise kinds of knowledge which cannot be protected by a patent, such as:

- discoveries, scientific theories or mathematical methods,
- aesthetic creations,
- schemes, rules or methods for performing mental acts, for playing games or for doing business, and computer programs, and
- information presentations.

Methods for surgical or therapeutic treatment or diagnostic methods, practised on humans or animals, are not to be regarded as inventions. This does not prevent patents being granted for products, including substances and compositions of substances, used for these purposes.

One of the fastest developing branch of patents is that of microbiological inventions. The general principle of most countries is that such inventions are patentable, but that patent protection is not possible for plant and animal varieties or essentially biological processes for the production of plants and animals. On 20 October 1988 the Commission of the European Communities issued a proposal for a Council Directive on the legal protection of biotechno-

1. Most issues discussed here were on the agenda of the 'Third International Civil Law Conference' in Stockholm on 10–14 June 1991, although not necessarily in the same sequence nor with the same content. The author's opinion may also deviate from the majority opinion of the conference.

logical inventions. There has been much discussion on the matter since then, and the praxis is not uniform. It seems that inventions relating to biotechnical processes in general are patentable but genetically engineered animals and plants are not.

Plant and animal varieties and biological processes for the production of plants or animals are hence only patentable in some countries, including Hungary.

b. Computer programs
Computer programs are not patentable. Parts thereof, which contain original steps due to a creative activity, are protected under copyright legislation in a number of countries.[2] The so-called routine softwares are not entitled to this protection. In principle it cannot be excluded that a program running a computer may come under patent protection, if the hardware itself is under such protection.

c. Semi-conductor chips
There was a long debate concerning the protection of semiconductor chips in the 1980s. Many specialists wanted to protect these worldwide under copyright law. Instead, a special protection was devised for them. The USA, most members of the European Union and most Nordic countries enacted special laws for the protection of semiconductor chip layouts. Hungary has enacted a similar special law too[3] according to which semiconductor chip layouts are protected, provided they are original.

d. Literary items (publications)
Technical, economic, commercial and financial information may appear in literary works or publications. In other words everything, which may form the subject of a transfer of technology, may also be published in a literary work. The question arises how far this is protected.

The literary work itself is under copyright protection (in Hungary under the protection of the already mentioned Act III of 1969), provided its contents constitute an original work. However, this means only that the author is protected against the unauthorised multiplication or edition of all or part of his work, against neglecting to mention his authorship, against distortion of the content, etc. He is not protected against the use of any of the knowledge contained therein by anybody. (It is quite another issue, whether a literary description is sufficient in itself to allow a real transfer of technology.)

2. In Hungary this is Act III of 1969.
3. Act XXXIX of 1991, with force from 1 Jan. 1992.

e. Know-how

As discussed in Chapter 1, know-how includes such technical and non-technical knowledge and experience, which cannot be protected otherwise (including surgical and therapeutic treatments and diagnostic methods). Know-how is an essential element of technology transfers, and some technological elements are provided with an additional, but more stable protection, since the patent protection is non-applicable.

On the other hand, know-how is not exclusive, hence if someone comes to the same knowledge and experience as somebody else before him, he may freely use and transfer it. This is different from the case of exclusive rights, which are in the exclusive property of their owner.

The law defends those intellectual creations which are not protected by separate provisions of law, but which may be used widely without having become a part of public knowledge. This relates to know-how as well. Due to the fact that know-how cannot be registered and hence has no exclusive protection, the protection given is against its illegal acquirement, use and publishing by third persons. (The prohibition of its publishing is important, because once it has been published, it becomes part of the public knowledge and no protection would be possible.)

f. Others

Utility models and designs are exclusive rights and may form the subject of a technology transfer.

The transfer of technology is not directly coupled to trade marks and trade names, but may sometimes encompass such markings. Trade marks are endowed with exclusive rights.

1.2. Infringements

Should one be infringed in his exclusive rights, he may ask for one of the following redresses:

- the ascertainment of the infringement by the courts,
- cessation of the infringement and the prohibition of further infringements,
- that the infringer make adequate amends, e.g. in the form of a public apology at infringer's expense,
- the discontinuance of the injurious situation, the restitution of the situation existing prior to the infringement, or the destruction of infringer's products or the elimination of their infringing nature,
- the seizure of the means by which the infringement was committed,
- damages according to the liability rules of the Civil Code.

Over and above this, the entitled person may also claim, within the frame of the protection of economic, technical and organisational knowledge, that the person unrightfully using it should share with him the gains achieved thereby. An infringement may also have some penal consequences.

The exclusive rights are bound to the jurisdiction of a certain country. Exclusive rights may be obtained in other countries too, and if the creation is protected there its owner may take the same measures and make the same or similar claims against an infringer. Such redresses are however not available for rights which cannot be registered and therefore are not exclusive ones. So e.g. the international protection of know-how is relatively weak, although it is protected by unfair competition laws in most countries. This is the reason why in the international transfer of technology the transferor secures its technology in the form of contracts or agreements. The means of this protection are either the so-called secrecy agreements, or the secrecy clauses of the know-how and the delivery of goods agreements.

2. SECRECY CLAUSES

The condition of secrecy is an organic part of know-how agreements, licence agreements which contain also a know-how component, technology transfers including engineering and delivery agreements for complete plants. That is to say: it plays a role in all kinds of technology transfers.

The 'secrecy clause' goes farther than the meaning conveyed by its simple reading. The essence of such secrecy clauses is on the one hand that the transferee has to keep the knowledge he received secret, i.e. he should not publish it, and on the other hand that he can use it only within the limits allowed and may not pass it to third parties. Characteristic of such clauses is the triple requirement of:

- the undertaking of the obligation of non-publication,
- the making use of it only with the limits permitted,
- the abstaining from the passing on of the information.

The requirement of *non-publication* (non-disclosure) needs no further comment.

The *use within the limits permitted* means that it has to be stipulated unambiguously in the other parts of the agreement, whether the permit allows the production and/or the sale of the product, in which territory the transferee has the right of production and in which that of sale.

In the case of know-how the permission for production is more or less a natural consequence of the agreement, but its place may not. Should the transferee have the right to sell the product, which is natural at least for the country where he produces it, it should be specified where the transferee has no right to sell (because e.g. another transfer has taken place for that country and the transferee of that country has the exclusive sales right for that territory), provided such stipulations are not contrary to anti-trust legislation prohibiting the incorporating of restricting conditions in the contract (*see* paragraph 4.3 of this chapter).

Should the technology be protected by a patent which is valid in more than one country, the transferor may curtail the transferee in respect of the countries where he may sell his goods (manufactured according to the transfer). It is however not legitimate to exclude from the right of sale a country where the patent has not been applied for. In cases where the transfer is taking place in the patent application stage, the question sometimes arises whether export can be prohibited to a third country where a patent is not yet in force. In our opinion, the export towards countries where the patent had been applied for can be prohibited, at least provisionally.

A further problem emerges in respect of the countries belonging to the European Common Market. It has not yet been settled whether export to another member country of the European Union can be prohibited by the licensor, even if there is a valid patent in that other member country.

Aside from the international aspect, a patent-based transfer may be limited to a certain part of the country (e.g. to the western half of the USA), or to certain claims of the patent, or to the production and/or the sale of certain volumes. Know-how agreements may also contain quantity limitations.

It is also not without interest whether the licence, i.e. the right of use, is an exclusive or a non-exclusive one, many contracts being of the latter type. A foreign partner may grant exclusive production and sales rights in the transfer agreement for a certain territory, say Hungary, but for other countries where a patent exists he may grant non-exclusive rights only, or – as already mentioned – none at all. Another possibility is that the transferee gets exclusive production rights for his country but only non-exclusive sales rights, so the transferor can also sell on that market.

The transferor may also grant exclusive production rights for a given country in the framework of a know-how agreement. This however merely means that he shall abstain from giving a right of use to others.

Some other kinds of limitation are also encountered. The *duration* for which the patent licence is given may be limited in the patent licence agreement. After its expiry the production and also the sales rights cease to exist. Should there be no such stipulation, it may be disputed whether the patentee has the right to take measures against the transferee of the licence during the period of existence of the patent (after expiration of the patent he has no such right) but after expiry of the agreement.

Another type of limitation deals with the *quantity.* Should a transferor deliver, in the framework of a patent licence agreement or a know-how agreement, a plant with a given capacity to transferee, he should also clarify in the agreement whether the transferee has the right to enlarge its capacity in the future, or whether he is obliged to keep the capacity at the initial level.

All these limitations have to be in harmony with the countervalue paid for the intellectual property rights. In case of a dispute, e.g. if the transferee deems that more rights are due to him under the agreement, the courts will investigate the question whether the service rendered and its countervalue have been balanced.

Under the secrecy clause the transferee undertakes the *abstaining from transmission* too. This means that the transferee cannot give any sublicence without the explicit permission of the transferor.

Some forms of transmission may however not be fully excluded, since in the course of the realisation of a plant other contractors may take part and thereby become at least partly aware of the knowledge received. The prohibition does not relate to this knowledge, and it may be necessary to explicitly stipulate that those persons who get acquainted with a part of the knowledge in this way are also obliged to keep it secret. In general such secrecy clauses stipulate that the employees of the transferee and all others who get acquainted with any part of the know-how during its transfer are obliged to keep it secret.

The above-mentioned triple obligation of secrecy does not relate to any knowledge which was verifiably in the possession of the transferee before the conclusion of the agreement. Such knowledge ought to be mentioned in the secrecy clause.

The secrecy requirement also does not pertain to that knowledge, which becomes publicly known in any other way than through the neglect of the transferee, nor to knowledge which transferee acquires lawfully from third parties.

If a relatively developed transferee possesses a lot of knowledge in the area of the future transfer, it is prudent for him to specify in writing all that was available to him before the transfer and to deposit it with a notary public. The risk of posterior disputes may be eliminated in this way.

The secrecy itself is normally binding for certain *period of time.* A period of seven to fourteen years is the general practice. After expiry of this period – according to our opinion – the transferee is released from its secrecy obligation, should there be no stipulation to the contrary. He may e.g. enlarge the capacity of his plant without payment to the transferor.

In the case of patent licence agreements the expiry of the obligation normally coincides with the expiry of the duration of the protective right of the last patent falling under the licence agreement.

A common clause is that the secrecy obligation survives the expiry of the agreement itself, since the agreement is often fulfilled earlier than the expiry of the secrecy obligation.

There are no clear-cut directives on how to deal with the secrecy issue if an agreement concerns a spread transfer, e.g. if the later transfer of new developments is included. It is usual that there are several secrecy obligations, each starting from the date of transfer of the corresponding knowledge. The duration may also begin with the last transfer of information, but the secrecy obligation is in such cases shorter, say 5–7 years.

The infringement of the secrecy clause has serious consequences, the foremost being the payment of damages.

In case of a know-how transfer, the publication of the know-how is the most harmful as it would undermine its essential character. Because of this, special care must be taken to check that any knowledge contained in a publication by the employees of the transferee is their original work and did not come from the transferor.

Using the transferred knowledge outside the framework laid down in the secrecy clause may, in case of infringement of patent rights in third countries, result in a patent infringement civil and penal lawsuit (e.g. if the goods are exported to a country which is excluded according to the agreement but where there is a valid patent).

The infringement of the prohibition to transmit secret information may also involve the payment of damages. If the third party to whom it was transmitted in violation of the contract publishes it, the default of the transferee is to be regarded as a grave one.

It has to be remarked, finally, that in some situations explicit secrecy agreements are concluded without any other deals. These cases shall be discussed under paragraph 2.1. of Chapter 4.

3. POSSIBLE CONSTRAINTS IN RESPECT OF EMPLOYEES

The employees on both sides, i.e. on the side of the transferor of the technology and on the side of the transferee, are held to secrecy. It is irrelevant whether the transferor grants a licence which he has developed himself, or has as a licensee received it from somebody else.

Some issues emerge automatically, such as:

- how the employer should oblige his employees to secrecy,
- whether the employees can be obliged to secrecy after the expiry of their employment,
- how to differentiate between the know-how of the employer, to be kept secret, and the personal know-how of the employee.

'The employees must keep their employers' secret know-how secret.' It does not matter, according to the report of the Conference of Stockholm (*see* note 1 above), whether the employer has transmitted the know-how to other companies, e.g. the employer has granted licences, or whether the employer has acquired the know-how from other enterprises. In some countries, e.g. Belgium, the obligation to keep the employers' know-how secret also lasts after the employment has come to an end, in others, e.g. Sweden, the obligation ends with the employment itself.

> 'Within certain limits it seems possible in all countries for an employer to strengthen the obligation to keep know-how secret by entering into secrecy agreements with its employees.'

In Hungary, the stipulations of paragraphs 1–3 of § 5 of Act LXXXVI of 1990 on the prohibition of unfair market behaviour are as follows:

> '(1) It is prohibited to acquire or to use a trade secret in an unfair way, or to communicate same unauthorised to others, or to publish it.
> (2) As an unfair way of acquisition of a trade secret is also to be regarded, if the trade secret has been acquired without permission of the entitled, with the cooperation of a person who is in a fiduciary relation or in a commercial connection with him.
> (3) In the performance of this Act
> *a.* trade secrets are all those facts, information, solutions or data in connection with an economic activity, to the secret keeping of which an appreciable interest of the entitled is coupled,
> *b.* a fiduciary relation is in particular the employment, another legal relation of work, or a membership relation,
> *c.* a business relation means the informative discussions and offers preceding the conclusion of an agreement too, even if no agreement is reached.'

It is obvious from the above-mentioned paragraphs of the Act, that the employees are obliged to secrecy. The law however does not declare whether the employees are obliged to secrecy after the end of their employment. Due to this fact it is advisable – at least in Hungary and in other countries with similar legislation – to incorporate into the employment agreement a secrecy clause, at least for those employees who might come into contact with trade secrets. In such secrecy agreements concluded with employees it must be specified, what kind of knowledge has to be kept secret by them, and for how long a time after the end of their employment.

In this respect we may mention the famous litigation case between the B.F. Goodrich company and one of its employees who left the company, Mr. Wohlgemuth.[4] The B.F. Goodrich company – in order to safeguard its know-how connected to the production of space clothing – started a lawsuit to bar the former chief of its space clothing production department from being employed in the similar department of its competitor. In the course of the lawsuit it was ascertained that although the know-how in question did not cover the whole procedure, it related to a series of individual part-solutions (e.g. to the production of the hard covering of the space helmet, the insertion of the visual aperture, the mounting of the helmet onto the space clothing, the production of the hard covering of the gloves), but the partial knowledges together practically gave the full (complex) knowledge of production of a certain type of space clothing.

4. Brooks, *New Yorker Magazine*, 11 January 1964, p. 37 ff.

The decision in the *B.F. Goodrich* versus *Wohlgemuth* case was that Mr. Wohlgemuth was allowed to take a job with the competitor, but was compelled to forget his knowledge acquired at B.F. Goodrich.

It is a serious problem, how the mixing of company know-how with the personal know-how of the employee can be prevented. Professional specialists are in the possession of certain knowledge due to their education, the so-called basic knowledge. They may however acquire special, individual knowledge (know-how) by working within a team, and the more teams in which they have worked, the more special knowledge they will probably have acquired. Personal know-how thus evolves and the specialist may sell it in the form of advice to other teams, or by accepting employment with another employer for a greater salary than his present employer is willing to give. Knowledge may be spread involuntarily too, because after a certain time a specialist may be convinced that the knowledge in question is his personal know-how and feel that he is entitled to use it as such.

The situation becomes especially difficult, if the employee gets acquainted with transferred know-how. A possible defence against diffusion by employees is that the transferor states unambiguously which part of his knowledge shall be kept secret, because it is unavoidable that otherwise the different knowledges mix up. It is however not advisable to stipulate that every bit of knowledge shall be kept secret, because by doing so, the transferor devalues that knowledge which really has to be kept secret.

Of course in the end all knowledge becomes part of the public domain, but it is not without interest, how long it can be kept secret, exploited economically and avoided that others should share it.

4. POSSIBLE LIMITATIONS

The freedom of contract is also valid in respect of technology transfer agreements. Thus there is no hindrance against concluding exclusive transfer agreements. There are also no restrictions in respect of the validity period of a transfer agreement. In some countries technology transfer is bound to special permits, but there may also be other types of limitations. This may affect the transfer itself, the transferee, or the freedom of contract.

4.1. The limitation of the transfer

The limitation of a transfer is mainly due to its subject. In several countries it is not legitimate to conclude a patent licence agreement based on an expired patent. Should however a patent, expired either because of the end of its protection period or because the periodic registration fees have not been paid – be accompanied by know-how, a know-how agreement may be concluded for the latter.

It should be remarked in this respect, that a technical transfer is the transfer of a certain technical knowledge, independent of the question whether it was patented or not.

A substantial number of limitations are motivated by purely national military-economic considerations. The limitations in respect of military technology are the most severe: if transfer thereof is not prohibited outright, it is at the least bound to permits. An example is the so-called COCOM lists, on the basis of which the transfer to certain countries of certain modern technical solutions and highly developed techniques which are militarily utilisable, is not permitted at all or is bound to permits. This list prohibited the export and transfer of certain types of technology to East European countries in the past, and remains valid for certain countries (e.g. Cuba and North Korea).

4.2. The possible limitations of transferee's activities

The transferee is limited in his activities, if the acquired technology is bound to permits. This situation no longer exists in Hungary, but continues to be the case in certain countries.

In certain countries, including Hungary, a transfer of technology in either direction must be notified to the authorities, first of all for statistical reasons. Such rules however cannot be regarded as limitations. It is further a rule in many countries, including Hungary, that transfer agreements have to be concluded in writing.

There are some countries where the import of technology is restricted because the payment of the transfer must be made in hard currency. In such cases an exchange permit or an import permit of the relevant authorities is necessary.

4.3. The possible limitation of the freedom of contract

> 'In some countries (e.g. Belgium), there exists no national *antitrust legislation* which has any impact, as far as transfer of technology is concerned. However, *in most countries the general anti-trust legislation is applicable to different kinds of transfer of technology agreements.* Besides the competition rules of the Rome Treaty further the secondary legislation, supplementing those rules, are of very great importance. In fact in many Western European countries the anti-trust legislation seems to contain the most far reaching limitations of the contracting parties' freedom to decide the contents of their agreement.'[5]

5. Report of the Stockholm Conference, *supra* note 1; italics added.

Anti-trust legislation is of course not limited to Western Europe. In Hungary the relevant law is the Unfair Market Behaviour Act (Act LXXXVI of 1990).

Let us discuss these anti-trust legislations in detail and see how they may limit the freedom of contract in the case of licence and know-how agreements. Although there are various legislations which may interfere with this freedom, it seems practical to discuss only the so-called anti-trust laws of the USA and those of the European Union, in so far as they may interfere with licence and know-how agreements. It has to be remarked in this respect that the anti-trust regulations of the USA are older, hence the legal practice in respect of those rules is quite stable. The relevant legislation of the European Union is more recent and in our opinion the legal practice has been changing throughout the years; it also has some bureaucratic features.

The practice of the anti-trust legislation of the USA differs in the case of a patent licence and that of a know-how agreement. This difference exists in the practice of the European Union as well.

The anti-trust laws which relate to licence and know-how agreements are the following:

- USA: sections 1 and 2 of the Sherman Act; section 3 and 7 of the Clayton Act; section 5(a)(1) of the Federal Trade Commission Act.
- European Union: §§ 85 and 86 of the Treaty of Rome.

These texts are reproduced in the appendix to this book.

4.3.1. Prohibited contractual stipulations

The literature enumerates six contractual stipulations which are prohibited both in the USA and the European Union (allowing for justified exceptions), and four further ones, which although not prohibited outright, would be annulled by a court in the case of litigation. The prohibited ones are:

1. A stipulation according to which the licensee is obliged to buy certain non-patented goods or services from the licensor. This is a so-called 'tie-in' clause. Exemptions were however made from this prohibition in cases where the licence is unusable without the licensors' raw material.
2. A stipulation which prohibits the licensee to sell products similar to those manufactured under the licence, but originating from competitors and not falling within the scope of the licence agreement. This is the so-called 'tie-out' clause.

3. The misuse of so-called 'package licensing',[6] according to which the licensee is obliged to take a licence for a group of patents, although he uses only a small part thereof. The 'package' *per se* is however not prohibited.
4. Payment of a licence fee over the full turnover, i.e. the obligation of the licensee to pay a licence fee on his global sales, even if part of the goods sold by him had nothing to do with the licence.
5. Payment of licence fees after expiry of the patent.[7]
6. A non-impugnment clause, i.e. a stipulation that the licensee may not open an annulment lawsuit against the patent which is the subject of the licence agreement.

Some types of stipulations may be *per se* void. There are four such cases:

1. a price agreement in respect of the goods to be sold,
2. a veto right of licensee in respect of other licensees (if the licence is non-exclusive),
3. the obligation of licensee to offer improvements and modifications achieved by him to licensor without payment or other compensation (so-called 'grant-back' clauses),
4. the limitation of the quantity to be produced.

In connection with the above-mentioned grant-back clauses, the problem especially seems to arise in the praxis of the USA if the development to be 'granted-back' is quite important, e.g. if it is more important and valuable than the basic solution. In the praxis of the European Union such grant-back clauses are allowed provided the transfer of new developments is bilateral and is made without payment.

4.3.2. Stipulations under the court's discretion

Some stipulations of an agreement may be annulled or upheld depending on the circumstances. We mention, first of all, *limitations in respect of the range of application* (such as technical fields, types of goods, etc.). In this respect the European Union seems to be more permissive since it allows a limitation of technical fields.

6. Under package licensing such a licence is to be understood, where the licensor grants a licence for many patents simultaneously, even for a package of up to 100 patents.
7. Should the subject of the licence agreement be more than a single patent (e.g. a package), the expiry of one or more patents does not prevent the payment of a licence fee (eventually a reduced one). The situation is similar if the subject of the agreement is a patent and know-how simultaneously. After expiry of the patent the payment of a fee for the know-how is not prohibited.

Also falling within the discretion of courts are *territorial limitations* (market limitations). In the USA, this eventual prohibition or permit seems to be the most important of all anti-trust regulations. According to US practice, a USA licensor may grant a licence for the whole country or only a part of it. The division of the world market however is contrary to the provisions of the Sherman Act. In case of know-how agreements such limitations are possible, provided they are limited in time.

In the European Union, unitary legislation for political reasons is placed above the patent practice of the member countries, and therefore market limitation within the European Union is allowed only with difficulty. This means that should a licensor within or even outside the European Union wish to grant a licence for one member country only, this limitation will probably be invalid, even if the ware is patented in the given country. For example, if a licensor prohibits goods produced under a German patent from being sold in France, this prohibition may be invalid even if he possesses a French patent for which no licence has been given. Such limitations may only be made if the licensor gets an exemption, e.g. on the basis that the total turnover of the companies which desire the limitation is below a certain value, and the export limitation is a temporary one.

As to know-how, the European Union rules are quite complicated. This is also the case for mixed contracts which contain both a patent licence and a know-how agreement. Attention is drawn, without going into details, to Commission Regulation No 556/89, issued on 30 November 1988,[8] on the application of Article 85(3) of the Treaty of Rome to certain categories of know-how licensing agreements.

In respect of markets outside the European Union, the situation is not clear. Decisions can be cited which allow market limitations, and others which prohibit it.

In the USA, cross-licensing and the forming of patent associations with more than two members is common. In itself this is not against the law, but to squeeze out a competitor from the market with the help of cross-licences and patent associations is prohibited.

In respect of the exclusivity of a licence the following could be mentioned:

- An exclusive licence is prohibited in the USA if the competition would considerably decrease thereby and a monopoly would be formed which contravenes the Clayton Act.
- The European Union tends to allow exclusive licences only for licensees whose turnover is relatively small (under 100 million ECU). In respect of know-how it does not acknowledge exclusivity.

8. Official Journal L 61 of 04.03.89, p. 1.

- In respect of stipulations whose validity depends on the judge's discretion, it is advisable in Europe for the parties to ask for prior approval of the competent authority of the European Union.

Anti-trust laws are applied in the USA first of all in cases where the licensee goes to court wishing to terminate his contract because of violation of the anti-trust legislation.

4.3.3. Hungarian decrees

The stipulations of Act No LXXXVI of 1990 on the prohibition of unfair market behaviour relate to licence and know-how agreements as well. There are as yet no court decisions concerning this matter.

5. INTERNATIONAL AGREEMENTS IN RESPECT OF TRANSFERS

There are no international agreements which deal explicitly with the transfer of technology itself, but there are a number of them which may affect the transfer. These are first of all international agreements which concern the protection of intellectual property and to which most countries adhere. The more important ones are the following:

- The Berne Convention for the Protection of Literary and Artistic Works. The latest revised text is the Brussels version (26 June 1948).
- The Universal Copyright Convention of Geneva Convention of 16 September 1952 (Paris). It is valid collaterally with the Berne Convention and has some identical, but no contradictory regulations. The signatories of the two Conventions are not all the same countries.
- The Paris Convention for the Protection of Industrial Property. This has quite a number of revised texts, the latest being the Stockholm revised text (14 July 1967). This Convention deals with the institution of priority (*see* paragraph 3 of Chapter 1).
- The European Patent Convention or EPC, concluded at Munich (5 October 1973). It opens up the possibility of filing patent applications at one place in one language with ensuing validity in more countries. The European Patent Office is in Munich.
- The Patent Cooperation Treaty (PTC) concluded on 19 June 1970 at Washington, similar to the EPC but with more members. Its central office is at Geneva.
- The Budapest Convention of (1977) concerning the deposition of micro-organisms for the purposes of patenting.
- The Paris Convention for the protection of plant varieties (1961).

- In connection with trade marks the Madrid Convention (as revised at Stockholm on 14 July 1967) offers the possibility of international registration of trade marks in those countries which signed the Convention.
- The Hague Convention of 1925 on the international deposition of designs.

The report of the Stockholm Conference on technology transfer of 1991 (*see supra* note 1) also mentions the following:

> 'As far as transfer of technology in a narrow sense is concerned, there seems to be no independent multilateral agreement. However, it ought to be mentioned that in 1980 the United Nations adopted a set of multilaterally agreed equitable principles and rules for the control of restrictive business practices. The principles are just recommendations. To a rather large extent the principles are focused on provisions which are common in international transfer of technology agreements of different kinds.'

6. SANCTIONS

The transfer of technology can be combined with various activities, with the delivery of documentation, with the delivery of machinery, with licence agreements, etc. All such contracts may include sanctions.

6.1. Breach of contract in general

Breaches of contract may be of as many kinds as there are obligations within the agreement. The most usual breaches are the following:

- late delivery (missing the delivery date), and
- faulty delivery, execution, etc. (defect of quality).

According to most civil codes, the entitled party may either request fulfilment in case of a delay in delivery, or if that does not serve his interests any more, may terminate the agreement. In both cases he may claim damages. Under Swiss law damages may be claimed independently from a penalty clause.

According to English law two groups of obligations exist. First, there are so-called warranties and in case of non-fulfilment the consequences are the payment of damages or other consequences as stipulated in the agreement. Second, there are so-called conditions on the non-fulfilment of which the entitled party may terminate the agreement. The delivery date may belong to either category, depending on the stipulations of the agreement.

French law also offers the choice between demanding fulfilment and termination of the agreement coupled with the claiming of damages.

Hungarian law (Civil Code § 300(1)) similarly gives a choice between fulfilment and termination plus damages. In the case of an agreement with a fixed term there is no need to prove the loss of interest in fulfilment in order to demand termination.

In the case of technology transfer agreements termination is rather rare.

The supplier has two defences:

- the unsuitable behaviour of the entitled (the transferee), and
- *force majeure* (act of God).

6.2. Sanctions in general

The expression 'sanctions' covers those contractual stipulations which deal with the non-fulfilment, or late fulfilment, by any of the parties of an obligation; thus, with failure, default and breach of contract. It is customary to deal with the sanctions not in a separate paragraph of the agreement, but at the relevant places (e.g. to specify sanctions for delayed delivery in the parts concerning the delivery of documentation and of goods).

Not every failure to fulfil comes within the punitive scope of the contract, because the law governing the agreement normally contains some stipulations in respect of failure or default of the parties.

The sanctions are generally the following:

- a penalty or liquidated damages,
- a price reduction,
- termination of the agreement (or a repudiation of thereof),
- general payment of damages.

6.3. Damages in general

Most civil codes contain rules on damages. We will take the example of Swiss substantive law.[9] The Swiss Civil Code states in section 41 that:

> 'He who causes to another wrongful damages, be it intentionally or by negligence, is liable to compensation.'

Section 100 says:

> 'An agreement according to which liability shall be excluded for unlawful intention or grave negligence is void.'

9. *Schweizerisches Obligationenrecht*, published by Orell Füssli Verlag, Zürich 1989.

Whether negligence is grave or light depends however upon judicial discretion.
Section 103 states:

'If the obliged party is in delay, he has to pay damages due to his delay (and is also liable for any accidents of chance). He can free himself from this liability if he proves that the delay did not happen due to any fault of his.'

The Hungarian Civil Code contains similar rules at § 299 (delay of the obliged party), § 314 (common rules for breach of contract), and §§ 339–344 (general rules for damages). Some main rules are:

'The obliged party is obliged to pay the damages of the entitled party except if he proves that he acted as could be expected from him in the given case to eliminate the delay.' (§ 299 (1))

'He who causes to another damage by a wrongful act is obliged to reimburse it. He is released from this responsibility, if he proves that he acted as could be expected from him in the given case.' (§ 339(1))

'The person suffering the damage is obliged to act to eliminate or reduce the damage, as could be generally expected from him in a given case. That fraction of the damage is not to be paid, which arose from the fact, that the damaged party did not fulfil this obligation.' (§ 340(1))

'A stipulation of a contract is void, if it restricts or excludes in advance the liability for the consequences ... of damages due to a wilful or grave negligence.' (§ 342(17))

6.4. Delay of delivery

In respect of a delay within reasonable limits (according to the nature of the case a few weeks or months) the most general sanction is a penalty or liquidated damages (the expression 'reasonable' could be replaced by the more meaningful 'tolerable'). If e.g. a party undertakes to prepare a study for the other party within 3 months, the tolerable delay is less (e.g. 2–3 weeks) than if a big plant is to be delivered (where the tolerable delay may be a few months).

We have mentioned that the sanction for an intolerable delay is a penalty or liquidated damages. These two concepts are not identical. The penalty does not prevent the party suffering the delay from claiming damages too, if he did in fact suffer damages due to the delay. Should he not have suffered such damages (e.g. the equipment was delivered late, but he could not have used it at the site since the foundations were not completed) he cannot claim damages and only the penalties are due to him, except if the agreement stipulates otherwise.

The penalty is transformed contractually into 'liquidated damages' if the obligation to pay penalties is coupled with the stipulation, that the payment of the penalty *excludes all other claims.* It must be remarked however that (as mentioned in paragraph 6.3 of this chapter) according to Swiss substantive law, which is often used for contracts, a stipulation is void if it excludes damages in case of grave negligence (in German: '*große Fahrlässigkeit*'). This is also the case in many other laws, including the Hungarian one.

It is however often hard to decide if a grave breach of contract has taken place. The courts would probably not regard as a grave breach of contract if a main contractor orders the equipment from a subcontractor belatedly, leading to a late delivery. It is certain, though, that if it could be proved that he intentionally or negligently did not order it and the delay is due to this fact, the courts would consider this a grave breach.

The 'absolute' payment obligation of a penalty (in case of a delay in delivery) may be overruled contractually. This is the case e.g. if the agreement provides that the supplier is only obliged to pay penalties if the delay has an effect on, for example, the scheduling of the erection. In this case, the supplier will not have to pay a penalty if the foundations are ready later than the equipment in question arrives, independent of the date.

Should the supplier undertake only approximate delivery terms, it becomes doubtful whether he has to pay penalties in case of delay, and from which date on. Imprecisely defined terms imply some tolerance (though it is hard to define how much), so smaller delays will surely not be penalised.

In case of liquidated damages the payment of extra damages is excluded. These damages, too, may be contractually overruled (e.g. they are to be paid only if the delay has had a real effect on the construction procedure) and they may further be made uncertain by the undertaking of only approximate terms.

A further general remark: should the delay be caused by the fault of the receiving party (e.g. the delivery is delayed because the supplier did not get forwarding instructions although he asked for them), the penalty is obviously not enforceable. This evident fact is sometimes forgotten, leading to superfluous arguments.

After these generalities, let us look at two kinds of sanctions which are common: those for the late delivery of documentation and those for the late delivery of goods. The first-mentioned one is of special importance for technology transfer agreements.

6.4.1. Penalty for delay in delivery of documentation

The *weekly* amount of penalty for the late delivery of documentation (under documentation is to be understood that which is produced in the course of the designing work, or which is readily available) is usually 0.5 per cent of the value of the documentation. It is usual to give a limit, e.g. 5–8 per cent of the value of the documentation. If the value of the documentation is not stated explicitly, then the basis is generally the total value of the agreement, but naturally at a lower rate, e.g. 0.05 per cent per week with a maximum of 0.5 to 1 per cent of the global price of the agreement.

The above shows that penalties have a very formal character. The late delivery of documentation may have a large adverse effect on the introduction of a technology or on the realisation period of an establishment, which is not reflected in the penalties. A higher penalty may be very difficult to be agreed upon, although in some cases a maximum of up to 2–3 per cent of the global price of the contract is agreed upon.

Another issue is the basis of the penalty. The total value of the documentation is only a fraction of the global price of the contract. This is especially so for the delivery of complete plants, where the value of each partial delivery is even lower. If the penalties are imposed per partial delivery (and based on the value thereof), they will be very low. However, the importance of the individual deliveries is unequal and the consequences of their belated receiving may be quite different. Sometimes a delay in the delivery of an important piece of documentation is penalised at another rate than that of an unimportant one, or the rate increases with the time of delay (e.g. the penalty is 0.5 per cent for the first week, 1 per cent for the second, 1.5 per cent for the third, etc.). Such progressive penalties only make sense if the accrued maximum is relatively high, that is 15–20 per cent of the total value of the documentation or 2 per cent of the global price of the contract, whichever is higher. It is to be noted that in the case of relatively unpackaged technology transfer agreements, the penalty for the belated delivery of documentation may easily reach 15–20 per cent of the global price.

Due to the difficulties encountered in defining the basis of the penalty, it may happen – mostly regarding the delivery of complete plants – that the penalty for late delivery of documentation is not a percentage but an absolute amount, which depends on the importance of the documentation and the length of the delay (e.g. a penalty of 100 USD/day for non-important documentation and 500 USD/day, rising later to 1000 USD/day, for important documentation).

In connection with delays in the delivery of documentation, the possibility has to be envisaged that the delay is not the fault of the designer but that of the transferee, e.g. if the latter does not furnish the necessary data in due time or modifies the already delivered data. In such cases the delivery terms of documentation must be modified in accordance with the delay in the delivery of data (not necessarily proportionally but case-by-case), a penalty being due only after the elapsing of the modified term.

The punctual delivery of the documentation is hard to enforce even with penalties in case the designer cannot fulfil his obligations due to e.g. a bottleneck in his own capacity, or if the terms agreed were unrealistic at the outset. In such cases the designer tries to evade penalties by *formally but not factually fulfilling his obligations in due time* (i.e. he supplies the prescribed documentation on time, but the contents need to be modified later in order to be usable). Such premeditated modifications can hardly be penalised, although the effective fulfilment happened with the arrival of the corrected documentation. Designers have an ample variety of excuses why modification was necessary, often that the transferee modified some initial data or requested an amendment.

Although the penalising of the belated delivery of documentation is fraught with difficulty, it is nevertheless practical to stipulate such penalties in order to at least stimulate to a certain extent the transferor-supplier to fulfil his obligations in due time.

6.4.2. Penalty for delay in delivery of goods

The penalty for the late delivery of goods is usually also 0.5 per cent per week with a maximum of 5–8 per cent of the full amount of the contract. Several questions may be raised in connection with this penalty too, such as whether the basis of the penalty is the value of the delayed goods (in case of phased delivery), the global value of the contract, or a mix thereof.

There are various alternative ways of penalisation in case of partial shipments. One method is for each delivery or a group of several deliveries to be separately penalised, another is that only the final deadline (delivery date) is penalised. An example of the first method is the stipulation that a penalty is to be paid in case of a delay in a partial shipment as follows:

- should the delay be less than 30 days, the basis of the penalty is the value of the goods in question,
- should the delay be within 31–60 days, the basis of the penalty is the value of that group of goods, or other larger unit, to which the shipment belongs,
- should the delay surpass 60 days, the basis of the penalty is the value of the plant unit to which the delayed shipment pertains or the global value of the contract.

The penalty rates may also differ, e.g. in the case of penalisation for a delay of up to 30 days, based on the value of the partial shipment, the rate for the first week may be 0.5 per cent, for the second week 1 per cent, etc. (Should the agreement not specify otherwise, a week is to be understood as a whole finished week.)

Another method of penalisation of partial shipments is through quarter-yearly itemisation of the delayed goods. In this case the possibility of rephasing is normally included in the agreement, but such rescheduling has to take place in due time (i.e. not after the agreed delivery date) and may not disturb the logical erection sequence (e.g. should equipment be delivered in more than one piece, then the bottom part of it is to be delivered first since without it the other parts cannot be mounted).

More than one penalisation method also exists when the penalty is bound to the final date of delivery. One way is to use as the basis for the penalty the global value of the contract. In other cases not the global value of the contract forms the basis of the penalty, but only the value of the delayed part; if the delay is a long one the basis may still become the global value of the contract. Again the penalty rates may vary.

It happens frequently – especially in the case of big units – that there is no penalty for the delayed delivery of small parts. The parties may consider that should the overwhelming majority of the big units be delivered on time, some smaller parts still missing can be delivered during the period of the erection without any disturbance at the site. In practice this is so formulated, that it is not to be regarded as a delay if the final delivery date is met for at least 94 or 96 per cent of all goods to be delivered and the missing parts do not hinder the erection (alternatively: 'if the missing parts are delivered within a predetermined extended time limit').

Some general remarks in respect of the penalisation of delivery terms are:

- should there be no other stipulation in the agreement, the terms are to be understood as relating to the place of performance;
- should the delay of delivery occur due to the fault of the transferee, it is not to be regarded as a 'delay' (in such cases the extension of the delivery terms is the logical solution, and penalisation starts only after said extended period);
- should the delay be due to *force majeure*, no penalties are to be paid.

6.4.3. Some other consequences of delays in delivery

In the above we have dealt with penalties and liquidated damages to be paid if the delivery terms are not met. The non-fulfilment of the delivery terms may however entail other sanctions too, such as the partial or full termination of the agreement.

Partial termination is not a possibility of principle only. It may happen that a party who otherwise fulfils its obligations according to the contract, cannot deliver one of the items in due time, e.g. because of a *force majeure* (for example, his subcontractor's factory burned down). In such cases the parties have to investigate whether it is quicker for the transferee to wait until the supplier can fulfil his obligations or for him to buy the missing item elsewhere. For this reason it is usual to stipulate in the agreement the partial termination or cancellation after a delay in the delivery of the relevant items of more than 3 or 6 months. As mentioned above, such a stipulation may also be made for delays due to *force majeure*, although in case of *force majeure* longer periods of grace are to be agreed before the right of termination or cancellation can be effected. The partial termination or cancellation is therefore equivalent to the cancellation of an item of the contract in question, due to practical considerations.

Full termination is stipulated in contracts in case of very long delays only. What 'long' means in this connection depends on the relevant circumstances. For the delivery of documentation or smaller objects this delay is shorter, e.g. 6 months. For the delivery of large plants it is normal to allow a longer delay (e.g. 9 months, and in case of *force majeure* over one year).

It is another issue, however, whether it is advantageous for the transferee to make use of this right. He will surely exercise it if the supplier did not deliver anything and he has no real hope of seeing the contract fulfilled. Should the fulfilment however be probable within a reasonable time, and he cannot acquire the requested knowledge or equipment from others before then, he will possibly not exercise his right to terminate but demand fulfilment. At the same time he may try to squeeze out the maximum in penalties and damages.

6.5. Sanctions for inferior quality

The elementary interest of the transferee is, that in case of (negative) deviations in quality the supplier should eliminate them. The sanctions partially serve this interest. The relevant sanctions try to regulate mainly those cases where the supplier is unable to eliminate the quality defects.

The types of sanction are penalties, price reductions, and others. Relatively small deviations and faults may be sanctioned by penalties or price reductions. For larger defects other legal consequences are usually prescribed.

The penalties to be paid in case of quality defects are basically of two kinds, namely:

a. a penalty for the delay caused by the time needed to correct the defect,
b. a penalty or price reduction depending on the seriousness of the defect.

The above measures may also be combined.

A penalty according to *a.* above is seldom prescribed in agreements, although it would be meaningful if the guaranteed qualities are not demonstrated in due time by the supplier. A penalty for the delay could be combined with the obligation to repair. In practice, though, suppliers do not often undertake such penalty obligations. When they are agreed, the penalty is imposed on a daily or weekly basis.

It is more often the case that a penalty-free extended time limit is allowed for repairs. The costs of the repairs are to be borne by the supplier and he has to pay for the costs of his experts at the site (which may add up to large sums in some cases). Should the supplier be unable to demonstrate the guaranteed qualities before the end of the extended time limit (that is, should he be unable to eliminate the quality defects), the obligation to pay penalties commences.

Various kinds of penalties may be stipulated. In case the defect leads to a lower productive capacity, it is usual to provide for penalties or price reductions which are proportional to the deviation in capacity, since the supplier has in effect supplied a smaller machine or plant. For example, in case of a 1 per cent reduction in capacity the penalty or price reduction may be 0.6 to 1 per cent.

If the quality of the produced goods is not up to the promised standard, the marketability is in danger. This would probably mean that the product has to be sold under its usual market price. The penalty or contracted price reduction may be calculated on this basis. If, however, the quality is so bad that the product becomes totally unacceptable for the market, the paying of penalties is insufficient.

Another defect may be that the energy, raw material, or other consumption requirements are higher than promised, leading to increased production costs. These costs can be calculated over a relevant period and converted into penalties.

The maximum amount of penalties may be limited in an agreement per parameter and also by an overall maximum. It is usual to limit all penalties to be paid under the agreement (i.e. the penalties for delayed deliveries and the penalties for quality defects) to a global maximum, whose value customarily varies between 7 to 15 per cent of the contract.

It must be remembered, however, that there are no hard and fast rules to cover any case. The main issue is whether the supplier may redeem quality defects by payment of penalties or a price reduction. When the defects are small ones, the supplier usually has the choice between repairing the faults or paying penalties (or granting a price reduction). It is not unusual to stipulate that, if the demonstration of quality fails even after the reparation has been made, the supplier is released of its further repair obligations only if he pays the maximum of penalties for the defect. This provision may be supplemented by a further rule stating that the aforementioned is only allowed if the deviation is not greater than a predetermined value.

Many variations exist for the case that the deviation from the prescribed parameters is not too large. What is meant by the expression 'not too large' depends on the case and should be clearly stated in the agreement itself. A deviation of 5–10 per cent in capacity, or a somewhat larger deviation in specific consumption values, is a realistic limit, but in some cases even 5 per cent seems to be too high.

If the supplier cannot remain within the agreed limit, more serious consequences have to be reckoned with. These may be separately dealt with per parameter, but it may also happen that the global amount of the penalties to be paid surpasses the maximum amount stated in the agreement. The latter case is regarded as one which was not foreseen by the parties, and it is therefore not disposed of in the agreement, leaving the decision to arbitration.

A great negative deviation from the guaranteed values is such a serious case which cannot be regulated in detail in a contract. When a so-called 'fiasco case' has developed, the contract may prescribe generalities only (e.g. a right of termination, payment of damages, etc.). The range of such cases is quite broad. It may happen that the negative deviation is just outside the limit, in which case payment of penalties can offset the damage, but it also happens that the equipment proves capable of producing only half the contracted capacity, or does not work at all. Such cases come within the domain of the arbitration courts.

Often the contract contains two limits for negative deviations, the first limit to be covered by penalties, the second entitling the transferee to have the equipment repaired by somebody else at the cost of the supplier, if such repair is at all feasible.

In case of a fiasco it is usual to make a so-called covering purchase, besides enforcing the right of termination, to alleviate the damages. The detailed analysis of fiasco cases goes beyond the scope of this book, but it has to be remarked that in such cases there is often grave negligence on the part of the transferor-supplier, for example when the transferor is supplying an unproven technology without explicitly mentioning this fact to transferee.

6.6. Special problems of licence and know-how agreements

We have already discussed the sanctions for delayed delivery. Documentation may be delivered also within the context of a licence or know-how agreement. These agreements usually contain substantive guarantees as well for the reaching of certain guaranteed parameters. Due to the fact that sometimes only documentation and no equipment is delivered, quite different sums constitute the security of the deal than in the case of a technology transfer combined with the delivery of equipment. The transferor may be liable up to even 50 per cent of the agreed licence fee or know-how remuneration.

Should the guaranteed values deviate from the guaranteed ones to a large extent, the possibility of termination may also be stipulated in such agreements, or the issue may be referred to an arbitration court. Of course this implies a large risk for the transferor of the technology or know-how, since he has already handed over his knowledge. Thus, even if the agreement states which documents have to be handed back in case of termination or repudiation of the agreement, the transferor is unable to prevent transferee (or its experts) from retaining the acquired knowledge. (This problem is present in all cases where the agreement is cancelled for any reason and the transferee is obliged to give back the received documents). This risk is the greater for the transferor because the extent to which the acquired knowledge may or will be used by the transferee is an unsettled issue.

In the case of a licence agreement, termination or cancellation presents different problems. While the contractual relation ceases to exist, the right of exclusivity remains. That is to say that even if the licensee is able to eliminate the technical fault which caused the termination or cancellation, and succeeds in putting the equipment in working order, he cannot use the patents or exclusive rights. Therefore, if he is able to have the equipment repaired, he will tend to be willing to agree with the licensor on a reduction of the licence fee or on some other benefits instead of exercising his right to terminate the agreement. It comes to termination or repudiation of the agreement only when the licensee considers the practical realisation of the invention hopeless.

Licence and know-how agreements may also contain some sanctions to protect the transferor's interests. These are:

- payment of damages in case of a breach of the secrecy obligation,
- termination, repudiation or cancellation of the agreement in case of non-payment of royalties.

It is difficult to estimate the damage which may be caused by a breach of the secrecy obligation when drawing up an agreement. The determination of the amount of damages to be payed by the transferee will generally be left to the discretion of an arbitration court.

In case of patent licence agreements, the non-payment of the running fees or royalties leads to termination or repudiation of the agreement, if this is stipulated in the agreement, or if it is not, to the initiation of cancellation proceedings. On the other hand, since in case of know-how agreements the situation may arise that the transferor has given the know-how out of his hands, the stipulation of a running royalty is risky. The normal procedure is to agree on a lump-sum remuneration (fee) and to allow its payment in instalments, against suitable security.

6.7. Securities

Sanctions alone do not suffice to guarantee that the supplier will fulfil his obligations properly and on time. This is why certain pecuniary and bank-like securities are often coupled to the sanctions. Such a security may be retention of a fraction of the price, a bank guarantee or a bill of exchange. It is similarly a normal condition, that the penalties are deductible from the outstanding payments of the transferee.

An automatic penalty deduction in case of a delay is not a recommended measure, since a formal overstepping of the delivery terms does not necessarily imply that the delay should be penalised (e.g. the delay may be due to some omission of the transferee, the transferee may have amended the delivery dates or communicated his acceptance of a certain delay, etc.). But it often happens that the bank participating in the transaction is not informed of the mitigating circumstances and applies the penalty automatically, the reparation of which takes a lot of energy and time.

A retained fraction of the price is to be paid out after the completion of the equipment or the plant. It may however happen, that the completion is protracted through transferee's fault. Consequently transferor-suppliers normally demand the payment of the full countervalue (price) within a certain period of time (the length of which, depending upon the character of the deal, varies between a few months to 1–2 years) reckoned from the last delivery date.

Performance securities (bank guarantees) serve the same aim of promising unconditional payment – without examination of the claim – to transferee. (However, should the transferee draw from the bank a certain amount without justified grounds, he has to bear all the consequences.) The duration of performance guarantees is limited; they either end at the quality acceptance (completion) or remain in force with a reduced amount up to the expiry of the mechanical guarantee. They cease to exist after a certain period of time, if the completion does not take place due to the fault of the transferee.

The order of magnitude of performance guarantees varies. It may be up to 15 per cent, to be reduced after completion to about 5 per cent at the most. For unpackaged licence and know-how agreements or engineering agreements, securities of 25 per cent cannot be regarded as unusual.

7. APPLICABLE LAW, ARBITRATION AND STATUTE OF LIMITATIONS

The applicable law, the arbitration court procedures and the statute of limitations are in a certain relationship to each other, because e.g. the statute of limitations is ruled by the applicable law, and so is the interpretation of the agreement by the arbitrators.

7.1. Applicable law

The applicable law is one of the most important stipulations in an agreement. The lack of such a provision may lead to an unclear situation.

Should one conclude a domestic agreement, the domestic law shall probably be applicable. Should one conclude however an agreement with foreigners – and transfer of technology agreements belong mainly to this category – the law of the transferor's country is often accepted, and in some cases that of a third country. In the Hungarian praxis, for example, the use of Swiss law is accepted since the Swiss substantive law (*Schweizerisches Obligationenrecht*) has advantages such as:

- it does not favour any of the parties and none of the parties comes into a favourable or less favourable position,
- the Swiss legislation is based on advanced production and commercial relations,
- besides having a well-detailed legislation, its juridical practice is also mature.

For technology transfer, sales, service and know-how agreements Swiss law contains few ambiguities. It may also be prescribed in the case of licence agreements, in which case, being the law of a third country, it prevails only as regards the stipulations of the agreement. In respect of patents or other registered rights (utility models, designs, trade marks) the patent or other relevant laws shall prevail of that country, in which the right in question is registered. Should more than one country be concerned, the law on industrial property rights of all these have to be taken into consideration. That is the reason why, in case of licence agreements and other transfer agreements which involve registered rights, it is not possible to exclude fully the *renvoi* to domestic laws.

7.2. Arbitration courts

The parties may agree that in cases of litigation the ordinary courts of the defendant's country shall be competent, but it is preferred that in case of dispute – should the parties be unable to come to an amiable resolution – not the ordinary courts, but an arbitration court shall decide. Should the parties however agree on the competence of an ordinary court, or should they maintain this possibility as an option (that is, the parties may elect either to bring the case before an ordinary court or before the arbitration court), they must also stipulate the court which shall be competent. This is ordinarily a court in the defendant's country.

In case of arbitration, various classifications of courts exist, such as according to the sphere of authority (whether the court has a general authority or only a special one within the specific country) and according to organisation (to which organisation it belongs). Arbitration courts may be general ones or belong to a certain professional union, to a stock exchange or a chamber. The so-called *ad hoc* courts of arbitration do not belong to any organisation.

Naturally the ordinary courts would remain competent in patent and industrial property litigations, since otherwise the decision of the arbitration court is hardly enforceable. For example, if the parties agree that the licensee cannot start a patent annulment procedure with the competent ordinary courts but must refer to the arbitration court, the decision of such an arbitration court – e.g. the granting of annulment – is valid only between the parties. Should the patentee disregard this decision, the licensee could hardly take any effective steps against him.

7.2.1. Ad hoc arbitration

In some foreign trade agreements *ad hoc* arbitration courts were previously much preferred, despite the cumbersome procedure for establishing them. Each of the parties first nominates an arbitrator. The chosen arbitrators then agree on nomination of a chairman. If they cannot agree on the person of the chairman, they pass on the duty of nomination to an institution which has previously consented to nominate a chairman. The previous consent of the institution selected to nominate the chairman is necessary since it would otherwise probably refuse the request on the grounds that it has nothing to do with the matter.

If an *ad hoc* arbitration court has been appointed, the court's statutes, place and working language must also be fixed in the agreement. Due to these inherent difficulties the practice has turned away from *ad hoc* arbitration courts and now favours the institutional ones.

7.2.2. Institutional arbitration

An institutional arbitration court is e.g. the arbitration court organised by the Chamber of Commerce of Paris. The advantage of such an arbitration court is, that the nomination procedure of the members and of the chairman, as well as the statutes, need not be stipulated in the agreement since they are included in the Rules of Arbitration of said Chamber.

A special advantage of the arbitration court of the Chamber of Commerce of Paris is, that the individual arbitrators and the chairman of the court may be chosen from a list of arbitrators set up by the member countries of the Chamber of Commerce. A party may propose as arbitrator a person not on the list, but the president of the arbitration court can be chosen only from the arbitrators on the list.

When an institutional arbitration court is provided in the agreement, it must be decided how many arbitrators there should be (e.g. 3), where the court should sit, and what should be its working language. Other stipulations are not necessary.

The decision of the court is announced in the language of the arbitration court's seat, and any of the parties may request a certified translation thereof. The decisions of the arbitration court are final and binding on the parties.

A new institutional arbitration court which may serve well in cases concerning transfer of technology is the one organised by WIPO (Geneva).

7.2.3. Enforcement of arbitration awards

A Convention for the enforcement of arbitration awards was concluded in New York in 1958, and many countries (about 60) have joined it. A European convention was concluded in 1961. All signatories to the Convention and those, who later join the Convention, recognise the validity of the arbitration awards of the corresponding court and are bound to enforce them. Should one of the parties request enforcement, respecting the formalities (enclosing a copy of the decision and of the agreement, or their translations), the recognition and enforcement can only be denied if the parties have no legal capacity, if the decision is null and void according to the law agreed by the parties or the law of the place where the decision took place, if the condemned party was not given the possibility of defence, or for similar formal reasons. Hungary joined the New York Convention in 1962.

7.3. Statute of limitations

The law regulates the statute of limitations normally according to what it relates to. So e.g. the agreed purchase price lapses after 2 years according to German law; according to English law the lapsing period is 6 years in the case of sales; while the period in French law is 10 years in respect of deals of businessmen with each other. According to paragraph (1) of § 324 of the Hungarian Civil Code: 'The lapsing time of claims is 5 years, should the provisions of law not stipulate otherwise.' According to paragraph (2) of the same section: 'With the lapsing of the main claim, the pending subsidiary claims lapse as well. The lapsing of independent subsidiary claims is not affected by the lapse of the main claim.'

Warranty claims normally lapse sooner. The statute of limitations for quality claims is relatively short: according to German and Austrian law only 6 months, and according to Swiss law one year *except if the seller agreed to a longer liability*. In respect of quality warranties the agreements themselves will thus prevail, hence the relevant provisions are to be formulated with the utmost care in technology transfer, licence, and/or know-how agreements.

4 | Problems of Contracting

1. TYPES OF CONTRACTS

According to paragraph (1) of § 198 of the Hungarian Civil Code:

> 'An obligation originates from an agreement to fulfil the service and an entitlement to demand said service.'

According to the character of the service there may be the following kinds of agreements:

- '*dare*' (= to give) obligations (e.g. contracts of sale, lease and tenancy, as well as services the aim of which is to transfer something);
- '*facere*' (= to do) obligations, dealing with the performance of management, a commission, etc., as well as the performance of an activity with the aim of financial gain (entrepreneurial contracts);
- '*praestare*' (= to live up to) obligations, namely obligations to accept responsibility (insurance, bank guarantees).

Most civil codes, including Swiss substantive law, enumerate the kinds of contracts. The Hungarian Civil Code itemises the types of contracts too, but technology transfer agreements do not fit in well. The kinds of contracts (agreements) enumerated in the Hungarian Civil Code, which can be related to technology transfers, are the following:

- sales agreements,
- entrepreneurial agreements,
- leasing agreements,
- tenancy agreements,
- commissionary agreements.

Agreements not itemised in most civil codes (nor in the Hungarian Civil Code) but relating to the transfer of technology are as follows:

- licence agreements for registered intellectual property rights (patent licence, trade mark licence, etc.),

- 'licence' agreements related to non-registered intellectual property rights (know-how agreements),
- engineering agreements,
- agreements relating to various services.

Moreover, so-called 'multi-subject' agreements contain a large variety of obligations.

Engineering agreements may be set up for the performance of various technical activities, from the preparing of a study to the designing activity at various levels, and from an assessment to technical consultancy. A multi-subject agreement may also have an engineering element. Such agreements may be contracts of sale, entrepreneurial contracts, or commissionary contracts. Agreements relating to various services may also be entrepreneurial ones (if they contain a liability for the result) or commissionary ones. Agreements of an individual character which could be put into the category 'others' may also appear in connection with a technology transfer as independent obligations or as a complement to other obligations (in the form of a collateral obligation), such as:

- agreements for the performance of a job or a reparation (of a commissionary or entrepreneurial character),
- cooperation agreements for the common performance of development (generally entrepreneurial agreements),
- agreements for the testing of equipment or a machine (generally commissionary agreements),
- a lease agreement concerning construction machinery or tools, which may either be an agreement of a simple lease or an agreement holding that the leased machine or equipment goes over to lessee's property with the expiry of the lease.

Above we also mentioned agreements which contain some elements of licensing. In practice, however, the agreements containing a licence are more complicated. The main subject of the mentioned licence agreements is namely the licence itself, but it may contain various collateral obligations which are relevant for judgement of the whole of the agreement. At the same time there are types of agreements where the licence is only a collateral obligation, and a faint one at that. It is hence practical to survey the kinds of agreements to be found which secure a licence. The following classification may be proposed:

a. Licence as the main subject of the agreement (contractual main obligation)

1. A licence for a patent, without collateral obligations (patent licence agreements).
2. An agreement concerning the transfer of know-how *per se* without any collateral service or delivery (know-how agreements).

3. An agreement concerning a patent licence together with the transfer of know-how, or *vice versa* (agreements with a mixed subject).
4. A mix of the above agreements, in which a collateral obligation is also stipulated (an accompanying service or delivery). Such agreements may be a patent licence, a know-how, or a mixed subject agreement having a collateral obligation, but the essence is that next to the licence for an intellectual property as the main subject of the agreement, a collateral obligation is also present.

b. Licence as a collateral subject (contractual collateral obligation)

1. Agreements for services (e.g. an engineering agreement) within which a patent or know-how is also transferred.
2. An agreement for the delivery of goods (agreement of sale for the delivery of equipment or a complete plant) in which a patent and/or know-how is also transferred.

c. *Special agreements* – e.g. an agreement for the foundation of an enterprise – in the frame of which one of the services is a patent licence and/or the transfer of know-how. Hereto belong also cooperation agreements in the course of which patents are licensed or know-how is transferred. (Such agreements shall be discussed in Chapter 5.)

While in the agreements under *a.* above it is possible to unambiguously judge whether their main subject is an intellectual creation (i.e. a patent licence or the transfer of know-how), in case of the agreements under *b.* and *c.* this is not necessarily the case. There it may not at all or only indirectly be determined whether their subject is (as a collateral obligation) also a patent licence and/or the transfer of know-how.

It is to be emphasised that even where a patent licence and/or the transfer of know-how is not the main subject of the agreement, the agreement has to stipulate the treatment of such intellectual property as well. It is not rare that when a licence forms the main subject of the agreement and is accompanied by some collateral service, the latter is drafted into a separate agreement. The drafting of a separate agreement also happens when it is the licence which is collateral.

1.1. Patent licence agreements

The subject of a patent licence agreement is a licence relating to registered industrial rights like a patent, a utility model or a design. A licence agreement for a utility model or a design does not differ fundamentally from a patent licence agreement, except in that a design licence agreement may be much simpler.

The grantee of a licence to a registered industrial right (the transferee) is entitled to what the *agreement explicitly permits.* This originates from the fact that in case of a registered industrial right the licensor (i.e. the transferor) has exclusive rights.

Patent licence agreements are non-typical and not categorised, according to continental law philosophy, but so-called *sui generis* (= peculiar) agreements,[1] which are ruled partly by the Patent Act and partly by the Civil Code. In Hungary §§ 17–20 of the Patent Act (Act No II of 1969)[2] is decisive next to the general dispositions of the Civil Code. The real subject of the patent licence agreement is namely strictly speaking solely the granting of a licence for one or more patents to the licensee (transferee).

To be valid a licence agreement has to contain the names of the parties, the subject of the agreement (enumeration of the patent or patents or other rights), the character of the licence (exclusive or non-exclusive), the duration, and the countervalue (price, fee and/or royalty). An agreement which contains the above is to be regarded as a valid one, provided it is put down in writing.

Patent licence agreements in today's practice however contain many more stipulations than the above, because the subject of the agreement is not only the transfer of the right of use (the licence), but also the transfer of the details of the technical solution contained in the patents. The same applies to utility model licence agreements. Since the design protects the aesthetic shape, the relevant licence agreements will probably contain less technical details.

As already mentioned in paragraph 12 of Chapter 1, a licence can be a simple (non-exclusive) or an exclusive one. This has to be mentioned in the agreement under any circumstances. It must also be stipulated in the agreement whether or not the transferee has the right to grant sub-licences.

1.2. Know-how agreements

Know-how agreements are also *sui generis* agreements, for which however the provisions of the patent legislation does not apply. Only the relevant parts of the Civil Code or similar provisions of law do. The difference between agreements concerning a registered right and know-how agreements is that the grantee (the transferee) of know-how is *not entitled to that which is explicitly prohibited by the agreement,* because there exist no exclusive rights, but only the transferor's market advantage.

One may also speak of a 'licence' in a know-how agreement, but due to the fact that it does not relate to registered industrial rights, it has no registration number and so one has to define in detail what technical or non-technical knowledge is the subject of the transfer. The type of licence has also to be stipulated: whether it is a non-exclusive or an exclusive one (exclusive however

1. Notwithstanding that the patent licence agreement is regarded as a *sui generis* agreement, it is often regarded – also in Hungarian domestic literature – as a special case of tenancy.
2. The new Patent Act (Act XXXIII of 1995) also contains stipulations in respect of the licence agreements (§§ 27–30).

only implies an abstention from offering the same licence to others). The countervalue of the use has also to be stated, as well as whether the transferee has the right to use the received know-how after expiry of the agreement, and whether he has the right to transmit or sub-license it to others. In any case a detailed description of the means and ways of the transfer of the technical or non-technical solution, as well as a secrecy clause will appear in the agreement.

1.3. Licence agreements with mixed subjects

These agreements relate to patent licences together with the transfer of know-how and form a special group of licence agreements. Their subject is on the one hand the licence of an exclusive right (namely: that of the patent) and on the other hand most often the transfer of a technical knowledge which is to be kept secret (the know-how), both together with possible collateral services or obligations. This duality means that careful contracting is necessary, the sensitive issues involved to be discussed now.

While the subject of a patent licence agreement is always the licensing of a patent right for exclusive or non-exclusive use, the subject of a know-how agreement is much harder to define. The main difference is that there is no secrecy obligation in connection with patents, while there always is in connection with know-how. The licence for a patent also in general lasts until the expiry of the patent rights, while in connection with know-how it is to be agreed for a certain time. Some stipulations of mixed subject agreements have hence to be harmonised in accordance with this duality, and others have to be defined separately in respect of the patent and of the know-how.

If the patent contained in a licence agreement with a mixed subject is annulled, does the agreement remain in force for the know-how and under what conditions? There is in principle no reason why the agreement cannot remain in force, but in practice many points of view have to be considered. The relative importance of the exclusive right in the deal must be weighed against that of the complementary know-how. If the know-how is the more important, the agreement will probably remain in force, and almost certainly if the transfer of improvements had also been agreed. If however the exclusive right dominates, the agreement shall probably cease to exist, or at least the countervalue should decrease since less is provided by the transferor. More precise statements cannot be made.

Another issue when the agreement remains in force after the annulment of the patent is its duration. This depends on the agreement itself and on a lot of incidental facts. In such cases the parties will try to come to an understanding, but should this be unsuccessful, the fate of the agreement will be determined by arbitrators.

Mixed contracts may however offer some advantages, and can come close to expressing the common interest of the contracting parties. For example, the duration of a patent licence agreement cannot be independent of the validity of the patent. That limitation may be circumvented if the agreement is declared to be a know-how agreement. Moreover, the payment of a large global sum for a patent licence may cause taxation problems in some countries, while know-how transfers do not. This may however vary from country to country and even be just the opposite in some countries.

To avoid these and similar problems the contracting parties may implicitly omit the reference to the one or the other subject of the agreement. Should the emphasis be on the transfer of know-how while the patent licence going with it is secondary importance, the latter may be omitted in the agreement and vice versa. (A remark however for the benefit of the transferee: if the agreement does not mention the patent licence, it is practical that simultaneously with the execution of the agreement, the patentee gives a 'free permit of use' of the patent in a separate declaration, or waives his right to commence a lawsuit of patent infringement against the transferee of the know-how).

If however the omission of one of the subjects of a mixed agreement is not feasible, the agreement should clearly contain both subjects, indicating too which has the precedence. Such agreements are to be drafted therefore with special care.

It must be kept in mind that patent licences can be entered into the patent register, but know-how 'licences' cannot be registered at all. For this reason the transferee will sometimes prefer a mixed agreement to a purely know-how contract.

A special case of mixed agreements is when the transferor owns patents in some countries, but not in the country where the transferee wishes to produce the patented solution. (E.g. this case occurs sometimes in Hungary, where the foreign patentee has no patent rights in Hungary but wishes to transfer a technical solution which relates to the invention to a Hungarian transferee, or just the opposite: a Hungarian company wishing to transfer a technical solution to countries where it has no patent rights). In such cases, should the agreement allow production and sale not only in countries where no patent exists but also in countries where the patent is valid, we may speak in one country of a know-how agreement, in the other of a patent licence agreement. This necessitates further differentiation.

A similar awkward situation may occur if the invention has been licensed already in its application phase, but the patent has not yet been granted in every country.

It may also happen that the deal was conceived originally as a know-how agreement, but because of an improvement of a part of the know-how or because of the patenting of the basic know-how itself, becomes transformed later into an agreement with a mixed subject. This possibility is to be considered in the original agreement.

The stipulations of such mixed agreements can be very varied indeed, and therefore cannot be discussed in all details.

1.4. Trade mark licence agreements

We mentioned trade marks in Chapter 1. For the use of a trade mark, too, one may conclude a licence agreement. This licence, however, differs from the aforementioned ones insofar as the licence concerns not a technical solution but the use of a special mark of goods. Due to this fact a specific problem is that the 'public interest' may interfere with the licence agreements. Having two or more companies use the same trade mark in the same country is undesirable, and so is the use of a trade mark which is well known in one country by someone else in another country, even if it is manufactured there with a licence.

Two main issues can be raised in connection with trade mark licences. The first is the exclusivity or non-exclusivity of the trade mark licence, the second, whether a special distinctive additional mark is to be used by the licensee.

Should the licensor and licensee of a trade mark be operative in the same country, the issue is how the two companies will appear on the market with a single trade mark. It is only feasible if the licensor strictly controls the *quality of the goods* of the licensee. Otherwise the licensor would undermine the market appreciation of his own goods and the 'goodwill' of his company. Having two companies with identical goods and identical trade marks on the same market is sensible only, where the quantity to be produced is quite large (a mass article which a single company is unable to produce in sufficient quantity to cover the market and therefore shares the production of through a technology transfer). This can also be the case if the country is so big (e.g. the USA, China) that a single producer is again unable to cover the whole of the market: the licensor may then wish to split the market – e.g. into western and eastern parts – with one or more other producers, whom he shall grant a trade mark licence and possibly provide with technology (through a technology transfer).

In the food, canned food, beer, soft drinks and tobacco industry another type of practice is current. A well-known (international) trade mark can have licensees in nearly every country, but the country of origin will be marked on the goods. This is accompanied by a strict quality control.

Market and trade mark sharing is however not often resorted to. Even in the case of a transfer of technology within the same country the tendency is for the transferee to wish to sell its product under its own trade mark and to try to enlarge its market share after a certain period.

It is more typical nowadays that firms which sell under the same trade mark within a country belong to the same trust. This simplifies the transfer of technology, the quality control and also the sharing of the market.

The case is different when the transfer is made to a company in another country. In such cases the use of the trade mark is allowed by the transferor not only when the transferee belongs to the same concern, but also if the transferor does not wish to export goods to the transferee's country.

It also happens often that only such trade mark use is allowed, where the goods are marked 'manufactured according to the system of X' (where X means a company with a good reputation). In other cases the transferor allows the use of its trade mark, but prescribes that a certain distinguishing mark should be added to the trade mark. Even in such cases the right of quality control is retained, and the use of the trade mark is allowed only as long as the quality of the goods manufactured by the transferee corresponds to the quality of the transferor's own goods.

If the transferor absolutely does not wish to produce the goods in the country of transferee or to export them there, an exclusive trade mark licence may be granted, assuming quality control of the goods produced. In this case the transferee shall probably be the sole (local) manufacturer of the goods under the licence. This situation may arise when it is not viable to export the goods in question to that country, because e.g. the transport costs are very high. A typical example of this setup is the soft drinks business.

The transferor is brought into difficulty if the transferee also wishes to export his goods to a country which is already supplied by the former. Here the only solution can be a sharing of the market. If both are delivering to the same market, care must be taken that the respective products should be recognisable as to their manufacturers. The trade mark licence is in such cases naturally a non-exclusive one for these countries – if the transferor allows the use of its trade mark at all – but this may lead to other difficulties, such as that the transferee may be compelled to use two trade marks for the same product.

Summing up: a trade mark licence lends a strong weapon to the transferee, so it should only be given if there is absolutely no interest in covering oneself the market concerned or if the licence remains within the same group of companies, and it seems advantageous to manufacture under the same trade mark everywhere.

2. PRE-CONTRACTUAL POSITION

In case of patent licence negotiations, it is an important requirement – directly following out of the nature of know-how – that the transfer of the rights should take place under the mutual confidence of the parties to the contract. In the case of know-how even the pre-contractual discussions are quite problematic, because the owner (transferor) has to give enough information to the future transferee to interest him in the transfer, but at the same time should not disclose details which he can use even in the absence of an agreement. This danger is smaller in the case of a patented invention, because the exclusive right granted by the patent does not allow the party asking for information to use it without the permission of the patentee (assuming that the patent is valid in the country where the possible transferee wants to produce). Should the discussion relate only to a patent application, the mentioned danger does exist, mainly in case a patent is not finally granted. If the patent is accompanied by know-how, which may be even more important than the patented invention itself, the negotiation technique is determined by the need to safeguard the non-registered know-how rights.

During the negotiations, the transferor's experts have to tread carefully. The natural instinct to disclose details in the heat of the debate, may lead to disadvantageous results for the future transferor. This danger may be averted by discussing only the technical final result and not the way of reaching it. Although the missing details will of course be included in the agreement, this often does not satisfy the transferee. The transferee may also need information which would allow him to make economic calculations, especially as to whether or not the intended agreement is profitable for him. This logical contradiction is very hard to resolve.

The situation is less severe if the parties are discussing the mutual exchange of know-how or if the parties do not discuss the simple transfer of know-how but couple it to a patent or to the delivery of goods (machinery or a complete plant). Here the situation of the transferor is more secure, e.g. because of his existing industrial property rights or because of the large volume of the business. Negotiations are also simplified when the know-how may only be realised by acquiring proper engineering and making an investment.

The situation is simpler if the party wishing to transfer the technology offers at the same time the delivery of machinery, i.e. he offers a complete plant or a turnkey job. In this case the transferee is assured that he will receive the required technology and also the equipment necessary for its realisation. Certain technical information should also be communicated during the pre-contractual phase, since transferee will naturally compare the offer with that of other potential transferors.

A different type of agreement may precede either the licence or the know-how contract, and also the offer for the delivery of a complete plant. For example, the potential transferee may first want to know whether the given technology may be realised by him economically. The usance is that in such cases a feasibility study is prepared first. Certain confidential data may slip into even such studies and it is also possible, that the exchange of confidential data is mutual, since the future transferee also has to supply data of a confidential nature (e.g. about his mines, about the site, his raw materials, etc.). The disclosure of such data may be delicate. Hence it is practical, that the parties should agree previously on the mutual confidential handling of the information received.

2.1. Secrecy agreements

We have already discussed the secrecy clauses of agreements in paragraph 2 of Chapter 3. Here we shall discuss secrecy agreements, to be concluded eventually as independent agreements.

Should the necessity of the disclosure of confidential information emerge during prior discussions, its illegal use, publication or transmission to third parties may be avoided by concluding a *preliminary agreement.* In such agreements the future transferee undertakes the obligation not to use, nor to transmit or publish the information received, particularly its confidential part. These secrecy agreements containing a unilateral obligation are common in practice.

Their term is either until the date of the later agreement – which supersedes the preliminary agreement – or until negotiations are halted. In this case the written part of the information received (descriptions, drawings, etc.) shall be handed back together with the eventual copies. The secrecy agreement may also provide that the confidential information received has to be kept secret for several years, failing which the party receiving the information shall be liable for damages.

The value of secrecy agreements is dubious in spite of all the above. The party receiving the information may later claim that the information received was publicly known and cannot be regarded as confidential, so the obligation undertaken has no juridical content. Should the obligation relate to detailed information connected to a patent application, and should the application later be rejected (e.g. for lack of novelty), this may also be an argument for contesting its juridical content.

The practice however shows that such agreements have a certain psychological value. Non-compliance with the agreement may give some short-term advantage to the obligee, but it may spoil its goodwill to a great extent.

A secrecy agreement may be supplemented with a previous payment obligation. Such a payment would be set off against the amounts to be paid under the final agreement. The setting of a low amount has however no restraining effect on a party undertaking the obligation *mala fide.* On the other hand it is not wise to stipulate payment of a considerable amount from the future buyer-transferee if he is *bona fide,* since this could refrain him from starting the preliminary discussions. In lieu of such 'earnest-money'-like deposits it may be better for the party undertaking the obligation to offer a bank guarantee to secure payment of damages in case of violation.

A secrecy agreement, in the frame of which both parties undertake a mutual obligation of secrecy, may simplify matters.

In Hungary there are some special provisions of law which are worth mentioning. Point c. of paragraph (3) of § 5 of Act LXXXVI of 1990 on Unfair Market Behaviour prohibits the illicit use, transmission and publication of all information, documents and data, which were handed over during the stage of tendering (*see* paragraph 2 of Chapter 3). This would make the conclusion of secrecy agreements (also with foreigners) unnecessary. However, this provision of law does not stipulate for how long the information received must be kept secret. So it may still serve the interest of the transferor to fix the duration of secrecy in a preliminary agreement.

If a party is interested in acquiring a technical solution of a know-how character and as a first step asks for a visit of transferor's plant, it is customary for the transferor to request from the experts visiting the plant or the laboratories the signing of a personal secrecy agreement. This obviously serves for the protection of the know-how. It also happens that the plant visit is bound to the payment of a certain sum, the so-called '*look and see fee*.'

3. THE USUAL CLAUSES OF LICENCE AND KNOW-HOW AGREEMENTS

We shall discuss the characteristic provisions of the licence and know-how agreements below. We will discuss separately the case where the licence and the know-how are only collateral subjects (e.g. they form part of an engineering or an agreement for the delivery of goods). Agreements for the delivery of complete plants will only be briefly treated, since the details are outside the scope of the present book.

3.1. The general aspects of licence and know-how agreements

The general aspects and the factual conditions of licence and know-how agreements are regulated in different ways by the various jurisdictions. Hence even the experts of the branch may not be aware of all the regulations, and even less of court practices. So we have to agree with Mr. P. Sebestyén[3] who gave the practical advice, that in the case of a licence agreement it is to be assumed that all the rules are working against our interests. Due to the fact that these rules are mainly dispositive ones, and therefore subject to modification in the agreement by the contracting parties, every question should be regulated from the point of view of that person on the side of whom we sit. Should we sit on the *export side* (Mr. Sebestyén gives an example of the so-called active licence agreements), it is pragmatic to consider that in the absence of any provision of the agreement to the contrary, the export of the patent:

- relates to the full technical area of the applicability of the invention,
- grants an exclusive licence,
- authorises the entitled to grant sub-licences,
- allows both production and sale,
- excludes the licensor from the realisation (sale),
- covers all geographical areas which are not expressly excluded,
- secures the registration of the licence,
- includes the obligation to deliver technical documentation, designs, drawings, etc.,
- makes the licensor liable on the basis of its legal warranty in case of actions of third parties,

3. P. Sebestyén, *Szabadalom kereskedelem* (Patent Trade), Budapest, 1963, p. 7 (in Hungarian only).

- makes the licensor liable for the novelty of the invention and for its technical feasibility, economic efficiency, competitiveness and quality,
- obliges the licensor to transfer later improvements free of charge, while the licensee is not reciprocally obliged to transfer its improvements,
- contains the risk that the licensee will claim that data relating to the invention were already known to him and that the patent was granted erroneously to the licensor, and subsequently attempt to get the same patent for himself.

All these issues must therefore be dealt with in the agreement and are to be regulated in a favourable way for the licensor (if we sit at that side).

The same issues come up if we sit on the side of the importer, the licensee, but now we must look upon it from another angle. In the case of export we assumed that in the absence of any provisions to the contrary the licence is an exclusive one, in the case of import we have to assume that it is a non-exclusive one. Where liability is to be supposed in case of export, we have to assume the opposite in case of import, i.e. unless liability is stipulated in the agreement the licensor shall have no warranty obligation whatsoever.

A similar approach is recommended for know-how agreements.

The above clearly shows that licence- and know-how agreements have to be drawn up with special care, and that one should stipulate all the important issues in detail to avoid divergent opinions and litigations later. An exception is when the issue is an unimportant one, or when the law governing the agreement is well known to the parties and they can rely on definite provisions of the law.

It is usual to have checklists as a help for the issues to be covered. Many such are known.[4]

It may occur that issues are left in the dark for tactical reasons, but reference to the dispositions of a certain law is recommended. These situations are however rather exceptional.

3.2. The structure of the contract

The practical structure of *patent licence agreements* is the following:

- the denomination of the parties;
- the intention of the contracting parties, the subject of the agreement (if besides the main subject some collateral obligations are present, these are to be stated explicitly);

4. *See* e.g. Stumpf, *Der Know-how Vertrag*, Anhang II: Verlags-gesellschaft Recht und Wirtschaft MBH, Heidelberg, 1970. *See also* Lontai, *A licenciaszerződések alapvető kérdései* (The Basic Issues of Licence Agreements), Közg. és Jogi Könyvkiadó, Budapest, 1978, p. 351 ff. (in Hungarian only).

- the basic obligations of the patentee (the licensor), such as:
  - the enumeration of those patents and/or patent applications for which a licence is granted by the licensor,
  - the character of the licence and determination of what the licensee is authorised to do,
  - whether the licence is registrable,
  - possible limitations in area, quantity or else in connection with the licence given,
  - the warranty of title of the licensor,
  - the possible technical or other (e.g. economic, marketing) guarantees of licensor,
  - an obligation for the protection and maintaining of the patent;
- the basic obligations of the licensee, namely:
  - the obligation of payment of the licence fee, its amount, the date and means of payment,
  - eventual further basic obligations of licensee (like the putting into practice of the patent, maintaining of the product quality, etc.);
- possible collateral obligations of the parties, e.g.
  - an obligation regarding mutual or one-way transmission of improvements, its duration, and treatment of eventual patentable inventions,
  - other collateral services (know-how, engineering, goods, etc.),
  - sub-licensing, the transferability of the licence,
  - action against infringers;
- the consequences of non-fulfilment of the licensor;
- the consequences of non-fulfilment of licensee;
- exemptions (*force majeure*, Act of God);
- general stipulations:
  - the applicable law,
  - litigation cases (the stipulation of an arbitration court),
  - the coming into force, or the dependence on the approval of the authorities,
  - the expiry or termination of the agreement,
  - the duration of the obligations;
- miscellaneous and closing stipulations.

The structure of *know-how agreements* differs somewhat from the previous ones, and may be e.g. as follows:

- the denomination of the parties;
- the intentions of the contracting parties, the subject of the agreement;
- the basic obligations of the party transferring the know-how:

  - · a description of the know-how to be transferred,
  - · the date of transfer,
  - · the way of transfer, the documentation to be provided, the training, etc.,
  - · eventual limitations in area, quantity or otherwise in connection with the transferred know-how,
  - · eventual further basic obligations (e.g. secrecy) of the transferor,
  - · the technical guarantees of the transferor;

- the basic obligations of the transferee of the know-how:

  - · the price (countervalue), the time and way of its payment,
  - · secrecy obligation,
  - · eventual further obligations of transferee (putting into practice, the guarantee of the quality of the product, etc.);

- possible collateral obligations, e.g.:

  - · an obligation regarding mutual or one-way transmission of improvements, its duration,
  - · a warranty of transferor in respect of the possible rights of third persons,
  - · other collateral services (engineering, goods, etc.);

- the consequences of the non-fulfilment of transferor;
- the consequences of the non-fulfilment of transferee;
- exemptions (*force majeure*, Act of God);
- general stipulations:

  - · the applicable law,
  - · litigation cases (the stipulation of an arbitration court),
  - · the coming into force, or the dependence on the approval of the authorities,
  - · the expiry or termination of the agreement,
  - · the duration of the obligations;

- miscellaneous and closing stipulations.

These typical structures of the agreements are of course not the only ones. For *mixed agreements* where the subject is a patent and know-how of approximately the same value (as opposed to the case where the know-how is of minor importance), the general structure could be refined in the sense that the basic obligations of the parties should incorporate those obligations which are characteristic both for patent licence and for know-how agreements. If a subject is of a minor importance only, it is advisable to draft the agreement according to the character of the main subject and include the collateral subject among the collateral obligations.

3.3. Contracting parties, their intention, subject of the agreement

There was a time in Hungary when only foreign trade companies could be contracting partners for concluding licence or know-how agreements, i.e. a technology transfer agreement with foreign companies. This may still be the case in countries with a so-called planned economy. Nowadays in Hungary practically any company can conclude a domestic or a foreign trade agreement.

A company may even act in the name of others abroad, e.g. an engineering or a trading company of a trust may act in the name of the trust. In such cases the acting company is acting either on commission or on the basis of an internal selling or cooperation agreement.

It is very important that licence and know-how agreements clearly state on the basis of what internal relations the parties are acting.

In simple patent licence agreements, the subject of the agreement is the licence itself. Should however the licensor provide other activities besides the granting of the licence – e.g. a designing or engineering activity, the supply of some parts or raw materials, taking part in the realisation of the subject of the patent, assistance at the putting into operation, etc. – these have to be clearly stated in the agreement. To avoid later disputes, such activities should be mentioned in the preamble of the agreement too. The intention, will and the accurate description of the subject of the agreement is very important for the interpretation of the agreement and may be decisive in the course of an arbitration procedure.

It is also advisable to state in the preamble of a know-how agreement – to avoid later disputes – the aim for which the agreement is being concluded, what its subject is, what character the given agreement has, what is to be transferred and how it is to be transferred.

A special problem of patent licence agreements may arise if the contracting party is not the patentee (e.g. a trading company is licensing patents abroad). In such cases it has to be mentioned either in the preamble of the agreement or at another adequate place (e.g. under the miscellaneous stipulations) in which capacity the contracting company is acting, otherwise it would be hard to assert certain rights, to register the licence for the company he was acting for, or to claim damages.

A similar problem may occur with know-how agreements if the entitled party is not identical to the party concluding the agreement (e.g. secrecy obligation, the granting of certain rights), although such problems generally occur only if the agreement is concluded by an intermediary on transferee's side and the rights are to be conferred latter.

3.4. Obligations of transferor

The licensor grants a licence basically for the exclusive right of use of a patent. The transferor of know-how transfers his experience and knowledge, and grants a licence for them within a predetermined scope. These two licences differ basically from each other and we therefore discuss the obligations of the licensor and the licensee and the grantor of a know-how and the grantee hereunder separately. The collateral obligations, guarantees and other contractual conditions shall be discussed together but only briefly.

3.4.1. Basic obligations of transferor in patent licence agreements

Due to the fact that the basic element of licence agreements is the licence granted for an exclusive right, the agreement should contain the *content* of this licence. First of all it is advisable to list all those patents and/or patent applications for which the licensor grants a licence. The subject of the licence may be one certain patent, or patents granted in various countries for the same invention – some of them maybe in the application stage –, but it may be that several patents or applications are granted or pending in a given technical branch in one or more countries.

The *character of the given licence* has to be determined (exclusive, with or without the right of licensor himself to use it, or non-exclusive). Should the subject of the agreement contain more patents and/or patent applications, it is possible that the agreement will be exclusive for only some of them but not for the others, or it may be exclusive in given countries and non-exclusive in others. For example, it is possible that a foreign licensor grants for its Hungarian patent an exclusive licence, but only a non-exclusive one or none at all for the patent in other countries.

An agreement should also contain a clear statement to what the agreement *authorises the licensee:* the manufacturing, the sale, or both. The content of the licence may vary here too according to countries. There are cases where the licensor sells on his own, so he grants only a licence for the production and buys the entire production from licensee, at least for a certain period. In other cases he grants both rights for some countries, but only the sale for others, and for certain countries he grants no licence at all. Thus a licensee may have e.g. a licence for production and sale in the domestic market, but only for sale in certain foreign countries; or a Hungarian company may grant an exclusive production and sales licence for Western Europe, but only a non-exclusive licence for sales overseas. The variation possibilities are infinite.

It has to be stipulated in the agreement whether the *registration* of the licence is allowed to the licensee.

The *limitations* in connection with licences may be of different kinds. These may be limitations in quantity (e.g. only a certain quantity may be produced yearly or may be sold), but they may also be territorial. Such limitations occur in most cases in case of non-exclusive licences. Territorial limitations are dubious when they relate to a country where the licensor has no patents at all. Due to the fact that one cannot legally limit the licensee in respect of such areas, such limitations can be replaced by other types of limitations, such as e.g. limitations in quantity or other suitable contractual constructions.

If the agreement contains no explicit limitations, that does not mean that they do not follow indirectly from the agreement. Should the licensor grant e.g. a licence for a patent in a single country and have a patent for the same invention in other countries too, the licence is valid only for the mentioned country. It then follows automatically from the agreement that the goods manufactured under the licence in the country it was granted for cannot be exported to countries where the licensor owns a patent for the same invention, even when the agreement does not mention this limitation. Should the licensee want to deliver the goods to other countries, this right has to be agreed separately in the agreement. This is also valid *vice versa:* should the licensor wish to limit the licensee, it is better to stipulate this in the agreement. It is however to be remarked, that the freedom of contracting is impaired in most countries, so such limitations may run into difficulties (*see* paragraph 4.3 of Chapter 3).

One thing in which the freedom of contracting is absolute, is the *duration.* Patent licence agreements may be concluded for any period falling within the patent validity (should there be more than one patents, up to the date of expiry of the last of them), and even for one-time use. It follows that in the agreement the duration of the licence has to be determined. (It is to be reiterated that a valid agreement cannot be concluded for an expired patent, because this may be used by everyone free of charge).

The issue of duration may appear in another form in licence agreements. Stipulations are quite usual according to which the licence is granted up to a certain time exclusively, but thereafter non exclusively. Instead of being time-bound, such transformation of the character of the licence may also come about after production of a certain quantity or, as we shall see later, after payment of certain fees, and all these may also be limited in time.

With respect to the *warranty of title* the licensor undertakes that should a third person hinder or limit the use of the licence, he shall hold licensee damage- and lawsuit-free. In Hungary the warranty of title of the licensor emerges from § 369 of the Hungarian Civil Code and is strengthened by paragraph (1) of § 18 of the Patent Act (or para. (1) of § 28 of the new Patent Act XXXIII of 1995). In some other jurisdictions the warranty of title does not follow automatically in every case, hence this issue has to be regulated too.

Claims may arise if a third party has a similar patent, i.e. if the one for which the licence was granted is really a dependent one, as it comes to the fore ulteriorly. The licensor's exclusive representative may also submit claims and request damages, saying that the grant of the exclusive licence made his representation rights suffer. Because of the risk of such and similar claims, licensees wish to stipulate a warranty of title while licensors wish to either avoid it or at least to limit it to a certain amount of money.

The *defence and maintenance* of patents is mostly the licensor's duty. Should annulment proceedings be filed against the patent, the licensor is obliged to defence. He may ask for the help of licensee in this action, but he has to bear the risk and costs of such proceedings himself. A stipulation to the contrary may lead to a reduced licence fee.

The payment of the yearly fees for the maintenance of the patent is also licensor's duty. There are cases, however, where the licensee takes over this duty. This issue has also to be clearly stipulated. Should the licensee take over the payment of the yearly fees for the maintenance of the patent, it should also be stated what liability the licensee has in this connection should he miss his duty and thereby cause the patent to expire.

3.4.2. Basic obligations of transferor in know-how agreements

Due to the fact that the nature of the know-how may not be determined as easily as in the case of a patent, it is important that the documents to be transferred should be listed, the details of the training should be determined and everything that pertains to how, when and what is to be transferred should be enumerated. The agreement itself often contains only the basic dispositions and stipulations, the details being settled a certain time after the conclusion of the agreement. This procedure is feasible if the transferor transfers more than a given single technical solution, but takes upon himself the liability to transfer all his knowledge and experience relating to a segment of industry to transferee. In such cases it would make no sense to drown the transferee in a great lot of documentation at a moment when he does not need more than a part of it. It is more practical for the experts of transferor and transferee to decide together – depending on the capabilities of transferor and the requirements of transferee – what should be the effective subject of the transfer, and how the transfer of the documents should be scheduled.

Naturally the content of the licence is also to be determined in the case of know-how agreements, namely to what it entitles the transferee of the know-how: to manufacturing, to sale or to both. The content of the licence may vary from country to country, and there are broader variations possible than in the case of the patent licence agreements. Due to the different nature of the licence, the range of the use is *limited* in know-how agreements.

Patent licence agreements are either exclusive or non-exclusive. Exclusivity is possible in the case of know-how agreements too, but it is not usual and would mean only that the transferor abstains from transferring the same know-how to others. Most know-how is transferred in a non-exclusive way.

It is impossible to guarantee that a third party will not work out and use or transfer a similar know-how, and so there is no such warranty in know-how agreements.

It happens often that the transferor limits the transferee in respect of the quantity to be produced or in respect of the geographical area, sometimes even in respect of the nature of the licence (manufacture or sale). An often used limitation is that the transferee may only use the know-how in a plant of a given size (and is obliged to pay further amounts if he wishes to enlarge his plant), or that he may only manufacture a limited volume of the product. A further limitation may be that production is allowed only in the home country or in certain countries, while sale is allowed everywhere or in certain geographical areas. Such limitations are linked to the price to be paid, but also to market conditions or to other circumstances (such as the fact that the transferor has sold already his know-how in a particular country).

Limitations in the know-how agreements are valid if properly drafted, although they may collide with anti-trust regulations to some extent (*see* paragraph 4.3 of Chapter 3).

The matter of duration is important also in know-how agreements, but somewhat differently than in patent licence agreements. A know-how agreement may be concluded in principle for an unlimited time (although due to the fact that the know-how may become obsolete, they may also be concluded for a given time only). An important question is whether the transferee has the right to use the know-how even after expiry of the agreement (hence without further payment). According to some views the transferee has the right to such use in the absence of provisions to the contrary, and also in the absence of provisions explicitly allowing it. It is however pragmatic if the agreement contains clear dispositions on this matter, the more if the real intention of the parties is not clear from the agreement.

Should know-how be transferred coupled to the delivery of a plant, it is clear that it may be used for an indefinite time within the confines of the plant. The duration of the licence cannot be disputed in this case, but the scope of the use, such as the enlargement of the plant, is another matter.

A further interesting issue – in case of certain agreed limitations – is whether the transferee has the right to modify the plant to increase production, using some other machinery or even other know-how. This would contradict the intention of the transferor to grant only a certain share of the market to transferee. Such cases may be settled through a posterior agreement between the parties, whereby the actual situation of transferor since the signing of the original agreement may have changed.

A basic obligation of the transferor is his *warranty* in respect of the possible rights of a third party. Within the frame of this warranty the transferor undertakes – unlimitedly or with certain limitations – that should any third party raise claims against transferee based on the latter's use of the transferred know-how, he warrants the exemption of transferee from lawsuits and damages. This warranty is similar to that mentioned for the patent licence agreements. It may however be of more importance than in the case of patent licences, since in the case of a patent the *ex officio* investigation of the novelty of the patent application often reveals whether any third party has rights which would hinder the utilisation of the patent (e.g. whether it is a dependent patent). In the case of know-how there is no such investigation of novelty, and so the transferor has to check himself whether any third party has a patent which would hinder the putting into practice of the know-how. Naturally not only patents may hinder the utilisation of the know-how, but other circumstances too (e.g. a third party raises a claim that the know-how originates from him and not from the transferor, and that the transferor obtained it illicitly; or an exclusive representative claims that due to the transfer of the know-how his rights are curtailed, etc.). Hence the transferee always requires the mentioned warranty, while the transferor strives to avoid it or at least to limit it to a certain amount of money.

There are sometimes stipulations which oblige the transferor to *keep secret* the know-how to be transferred, in the interest of the transferee. This is analogous to the obligation of the maintenance of patents.

3.5. Obligations of transferee

The licensee's and the transferee's basic obligation is to pay for the rights and/or knowledge he received. It is questionable whether this payment is to be regarded as a licence fee in all know-how agreements, since in the case of an aggregate sum payment the transferee does not pay a 'licence fee.' Since there are also other differences in respect of payments and of some other basic obligations, we shall discuss the obligations separately for licence and know-how agreements.

3.5.1. Basic obligations of transferee in patent licence agreements

The *licence fee,* its terms of payment and the way of paying it has to be regulated in detail in the agreement. This is of basic importance. The licence fee may be an aggregate sum, a non-recurrent expense, but this is relatively rare with patent licences. An advance payment plus a fee which depends on the quantity of produced units per year (a so-called royalty) is the usual solution. The licence fee may be a digressive one.

It is also usual to fix a *minimal licence fee,* which has to be paid by the licensee independently of the quantity produced per year, even if licensee does not make use of the licence and his production is nil. This minimal licence fee covers the royalty for a certain quantity and if the licensee produces more than that, he is obliged to pay more.

A maximum licence fee may also be provided, which means that the licensee is not obliged to pay more than this sum even if he produces more. This may be fixed per year and also as an absolute overall accumulated sum. After reaching this maximum, the licensee may use the licence free of charge.

The amount of the licence fee depends on the rights received and may also depend on the collateral services. For example, the value of a non-exclusive licence is lower than that of an exclusive one. Should production experience and/or know-how also be transferred, this will make the licence fee higher than it would be for a mere licence.

The fixing of a minimal licence fee is usual with exclusive licences, since the licensor is not allowed to give a licence to anyone else and will receive nothing should the licensee not use the licence.

It is also not unusual for the parties to an exclusive licence to agree that the licensee shall pay minimum licence fees but will, after the lapse of a certain period (3 or 5 years), have a choice: either to stick to the exclusive licence and pay the minimum royalties even if these are higher than warranted by his turnover, or to have the exclusive licence changed into a non-exclusive one and be exempted from the payment of the minimum royalties. The underlying philosophy here is that the licensee wants to be in sole possession of the licence during the first years in order to exploit the licence to a certain extent, master the production and establish himself on the market; later this exclusivity may be less important, or its price may be too high.

The number of possibilities for establishing the basis of royalty payments are simply countless. If the licensee produces only at a single site but sells (exports) to a number of markets, payment is often made based on production. Should he however produce at more than one site and sell in various countries, he may have to pay royalties based on his total production, or to pay a licence fee based only on the production at the main site but at a higher rate to compensate for the production at the other, subsidiary sites.

The agreement should contain very accurately:

- in what currency (or currencies) the licensee must pay the fee (royalty);
- when and on what basis the parties will do their accounting between them, and how long after calculating the fee to be paid the payment becomes due (e.g. the accountancy containing the turnover data of a year will be performed by February of the next year, and the payment shall be effected within 30 days from this date);
- the manner of the remittance and the receiver of the licence fees (royalties);
- whether the licensee is obliged to give certain securities and of what kind (e.g. a bank guarantee);
- what shall be the consequences of his non-payment in due time (e.g. there is a short period of grace for the payment, but interest is due thereafter; the possibility of suspension of the licence rights in case of a long delay in payment and of termination in case of recurrence).

Among the eventual obligations of the licensee one must emphasise the *compulsion of the 'working' (using) of the patent.* This obligation has a dual purpose. It serves on the one hand to ensure that the licensor will get a certain income (which can be achieved otherwise too, by stipulating a minimum licence fee). It is on the other hand connected with the patent practice in most countries, according to which the patent has to be put into practice (i.e. has to 'work) within a certain period of time, failing which the granting of a compulsory licence to a third party is possible, notwithstanding the given licence.

Another main obligation is *keeping the quality* of the goods produced under the licence at a certain level. Such clauses are often resorted to if the licensee mentions on the goods produced or in the sales literature, that the product was produced on the basis of a licence, or if it may be otherwise deduced from the product that it was produced under a licence of the licensor. The situation is even more touchy if the patent licence is accompanied by a trade mark licence (*see* paragraph 1.4 of this chapter). In such cases the licensor has the right to control the quality of the product, and non-adherence to the quality requirements may involve sanctions. (It is here to be remarked that, according to § 7 of the Hungarian Act No LXXXVI of 1990 on the Prohibition of Unfair Market Behaviour, if in a licence agreement there is no explicit permission given thereto, the licensee may not make reference to the licensor or the licence either on the goods or in his sales literature.)

It has been usual to include a stipulation that the *licensee has no right to commence an annulment procedure* against the patent for which he received a licence. The validity of such stipulations may however be doubtful due to the anti-trust laws of various countries.

3.5.2. Basic obligations of transferee in know-how agreements

The *countervalue* (price) of the know-how, the date of the payment and the way of payment are basic components of the relevant agreements, hence their regulation is important. The price may have a similar structure as in the case of licence agreements (e.g. a down payment and a sum which depends on the production). In the case of know-how the countervalue is usually fixed as an *aggregated sum*, although this amount may be paid also in instalments. (Behind this rule lies the fear that third parties may also work out and publish the same know-how. If this happens the transferee may justifiably ask for a reduction of the countervalue, or may deny any payment.)

The payment has to be in conformity with the duration of the agreement, with its probable prolongation or extension, with the transfer of improvements, etc. It may happen that the transferor requests a lump sum to be paid immediately after the transfer, whilst for the transfer of improvements a further yearly amount is to be paid, as laid down in the agreement.

Sometimes the transferee does not pay in cash but with goods produced according to the transferred know-how (or at least partly with it). Eventually the communication of the transferee's own know-how to the transferor may also form the countervalue (or part of it).

The payment obligation may be diminished – this can be stipulated in the agreement – or fully cancelled, when the know-how loses its secrecy. Such loss in value or in the interest of the transferee may also happen as a consequence of a new technical development.

The currency of the payment, its time and place, etc. must also be regulated in the agreement.

Another basic obligation of the transferee is the obligation of *secrecy*. This is a main element of all know-how agreements.

It is useful to determine what part of the information (knowledge, data, documents) should be kept secret. No such information has to be kept secret, which was publicly known at the moment of the transfer (which had been published earlier); which came later to be publicly known without the fault of the transferee; and of which it can be proven that it was in the possession of the transferee prior to the transfer. The transferee has the obligation to handle the information received confidentially and to prescribe the same to his employees; he may only use it for the purposes fixed in the agreement; and he may not publish it.

The transferee may come into possession of the know-how not only through documents, but also through training, which is the subject of many know-how agreements. Should a third party use this know-how which was acquired by training, this would infringe transferors' rights in the same way as the illicit use of documents. For the reparation of his injury the transferor may naturally claim damages. In practice a suspicion of illicit use may emerge when an expert in transferee's employ leaves his job after receiving the training and takes an employment at a competing firm. The transferee may defend himself against this risk by including relevant provisions in the employment contracts of the persons participating in the training, but it is better and more expedient if he requests the transferor of the know-how to oblige the person taking part in the training to sign a written declaration of secrecy.

The obligation to *keep the quality* at a certain level may also occur in know-how agreements although it is scarcer here, except when the transfer of know-how is bound to a trade-mark licence (*see* the previous paragraph of this chapter).

3.6. Collateral obligations

A series of collateral obligations exist both in patent licence and know-how agreements. We limit ourselves to a discussion of only a few of the most characteristic provisions, such as:

- the obligation to transfer improvements;
- services and deliveries of goods;
- the further transfer of the received licence (sub-licensing); and
- steps to be taken against illicit users.

3.6.1. The transfer of improvements

Both in the case of patent licence and know-how agreements the transfer of improvements is a frequent stipulation. The contractual provisions concerning the transfer of improvements or developments are varied and may also differ to a certain extent depending on whether a patent licence or a transfer of know-how is the subject of the agreement. Otherwise they have many common features.

In some kinds of know-how agreements it is usual to stipulate that the transferor is obliged to transfer the improvements up to a certain time (e.g. from contracting up to a period of 5 years) and the countervalue in the agreement includes such posterior transfers too. Another type of know-how agreements is that where the transferee receives the existing technical knowledge – the so-called backward information – at the time of contracting or shortly thereafter and at a predetermined time also receives the so-called forward information. In such cases the transferee pays a lump-sum fee for the backward information, and yearly royalties for the forward information. Such an agreement is preferable when a group of similar products is concerned, with the advantage for transferee that he does not have to employ development personnel since he receives the developments regularly.

In the case of patent licence agreements it is rather usual that the licensor transfers smaller improvements, falling into the scope of the licensed patent, free of charge. Their price is contained in the licence fee and in the royalties.

Both types of agreement are into rough sailing if essential technical improvements are realised. The licensor or the transferor will be reluctant to consider these as 'improvements' he are obliged to transfer. It is really quite difficult to define in an agreement beforehand what falls within the scope of improvements which must be transferred, and what does not. It is certainly not easy to declare a *new* patent as an improvement falling under the scope of a licence agreement (although it often is). It is usual to arrange this matter in such a way, that independent new patents do not fall under the improvements to be transferred. Nor do new solutions which lead to radical changes in the branch of industry, even if they are not patented.

It is also usual that in case of such improvements or changes the licensor or the transferor has the obligation to notify the other party and offer them for licensing to the licensee or transferee. The latter may get a licence or a transfer against a reasonable surplus price and thus come into possession of the result of the improvement.

New developments may however be of such importance, that the licence or the technical solution becomes practically obsolete. The parties may refer to that possibility in their agreement (if an essence of the agreement is the transfer of forward information), stipulating that if the interest of the transferee has been totally or in a major part lost because of a new development, the parties shall reconsider the contract. The reconsideration may lead to the conclusion of a new licence or know-how agreement, cancelling the old one, or to its retention with some modifications such as an increase of the original licence fee and the transfer of the new developments.

It would be hard to come to a more accurate regulation of future developments in years ahead than the above, since it is impossible to foresee whether an important improvement shall be found and with what consequences. One can only lay down that there shall be a reasonable or fair countervalue for such improvements, or that the price will be less than what third parties would be willing to pay, or that it will be a certain fraction of the market value.

A possible obligation in connection with the transfer of improvements is reciprocity. This means that not only the licensor or transferor is obliged to transfer his improvements, but the licensee or the transferee is too. This is called a grant-back clause. The grant-back clause is cumbersome when the licensee or the transferee achieves a patentable improvement. It is therefore usual to stipulate that improvements of minor importance shall be transferred to each other mutually and free of charge, while patentable ones shall also be transferred mutually but against payment. In this situation it must be decided whether the transferee of know-how is obliged to apply for a patent on an improvement elaborated by him, and in which countries; and should he not wish to apply for a patent, whether the licensor may apply in his stead. Also to be fixed is who will bear the costs of the patent application, to whom the eventual licence fees resulting from such patented improvements belong, etc.

A further issue in the case of an improvement patent is whether the right of use (the licence) is an exclusive or a non-exclusive one, independently of how this was regulated for the original licence.

If the improvement is a common one, the question arises in whose name it should be patented. It may be patented either by the transferor or by the transferee (with a silent agreement on each one's share), or be declared as a common patent with a declaration of each party's share.

Due to the fact that these and similar variations are almost inexhaustible, it is preferable to stipulate in the agreement only principles on the basis of which further agreement can be reached should the actual case arise.

One last point: should the obligation relate also to the transfer of patentable improvements (either unidirectionally or mutually), the scope of the agreement could be enlarged for the future and its duration extended. In the past it was usual to stipulate that only the licensee or transferee had to report his improvements and developments, the licensor or transferor having no such duty. This simplified the matter to a certain extent, but nowadays such stipulations are unpopular due to the anti-trust regulations all over the world.

3.6.2. Deliveries and services

The licensor is obliged to allow the use of a technology when granting a licence in licence agreements, but is not obliged – and often is not even able – to transfer the technical details, experience and know-how, and to render services in the realisation of the production such as engineering or technical assistance. Therefore, if the licensee wishes to receive such services, he has to stipulate these separately in the agreement.

The case is somewhat different in know-how agreements. Except in the case of agreements concerning a simple knowledge or experience or a single formula, it is nearly unavoidable that the transferor of the know-how will render various services for the sake of the realisation.

Such deliveries and services may be *optional* ones only: the licensor or the transferor of know-how undertakes a unilateral obligation to render, if the licensee or transferee so wishes, the deliveries or the services fixed in the agreement. In other cases it is the supplier of the licence or know-how, the transferor, who insists on certain deliveries or services, because he considers them necessary for the putting into practice of the technical knowledge, and therefore does not wish to undertake any guarantee if they are refused (*see* paragraph 3.7 of this chapter).

Various services may form a part of the licence or know-how agreement (as collateral obligations). It is also possible that the licensor or the transferor of the know-how concludes a separate agreement for them. Such services, in whatever way they are coupled to the main obligation, may basically be engineering activities, technical assistance or commercial assistance.

In the following we shall discuss the most frequent contractual stipulations for this type of services, with the remark that these can either form a part of licence and know-how agreements or can be drawn up as separate agreements.

a. The typical stipulations of engineering agreements

Since Swiss substantive law (*Obligationenrecht*) does not contain a separate category for a projecting (designing) contract, we refer to the Hungarian Civil Code, §§ 408–411 of which deals with the projecting contracts as a type of entrepreneurial contracts. It states:

> '§ 408
>
> On the basis of a projecting contract the entrepreneur obliges itself to perform a technical-economic designing work, the customer to take delivery and to pay a fee.
>
> § 409
>
> (1) The parties may agree in the contract to execute the technical-economic projecting work as a step by step service.
>
> (2) The parties may stipulate in the contract the limitation of the liability for damages, if the entrepreneur undertakes the presenting of a domestically not well known or not yet used technical-economic solution.
>
> (3) The customer may use the design only for the purpose appearing in the contract, and may not publish it.
>
> § 410
>
> (1) The commencement of the lapse of warranty rights enforceable due to a fault of the design starts at the moment of the fulfilment of the execution of the service. A provision of law may however regulate differently.

(2) The entrepreneur warrants that third parties do not have any rights, which would hinder or limit the realisation of the design. In respect of this warranty the rules for the warranty of the seller of property shall prevail as suitable.

§ 411
(1) Should one of the parties be an economic organisation, the general rules for entrepreneurial contracts shall prevail.
(2) Should the entrepreneur be an economic organisation, the parties may agree on the usage of the stipulations for the projecting contracts.
(3) Details of the projecting contracts shall be contained in separate rules.'

There is a government order in Hungary dealing with projecting contracts.[5] It is built up of the following parts:

- the contracting,
- the cooperation of the parties,
- the supervision of up-to-dateness,
- breach of contract,
- the general use of the rules of the entrepreneurial contracts.

There are some decisions of the Hungarian Supreme Court as well dealing with the projecting activity. One of them (No 197/1989) states that should the realisation of the project be impossible for the contracted sum, the supplier of the service has performed faultily. Another decision (No 334/1986) states that 'that project is faulty which envisages an uneconomic realisation, and the designer has to warrant for it. Hence it follows, that it may not claim an extra fee for the revision and redesigning the project into an economic one.' (In the given case the tracing of a sewage channel was found to be faulty because a shorter one could have been designed). Another decision is open to critique (No 281/1988). It states that 'the entrepreneur is not exempted from the consequences of the faulty fulfilment on the basis, that the detailed design did not contain the way of fastening.' This would mean however that the execution of the detailed design falls within the expertise of the contractor. In our opinion the detailed designs have to show such details too; only the basic designs do not need to.

Let us, after this introduction, discuss the engineering agreements. The typical stipulations of these agreements relate to:

- the scope and nature of the engineering to be executed,
- the sharing of the activity between the parties,
- plan coordination questions (especially in the case of task sharing),
- amendments and modifications,

5. §§ 83–89 of Order No 7/1978/II.1.MT (Council of Ministers).

– the documents to be delivered, terms, the issues of delivery,
– the question of standards,
– the manner of accounting the activity.

In respect of the scope and nature of the engineering it is not enough to simply fix what type of engineering has to be done, (e.g. a basic or a detailed one), but it has also to be laid down what sort of designs the supplier of the service produces in what detail, what sort of documents he delivers and when he delivers them. It is often necessary to have just a little bit more or less than what is generally understood under basic engineering, e.g. it may be necessary to have greater detail in respect of a specific item, but not for the rest.

Sometimes a basic engineering only is wanted for a part of the designs and a detailed engineering for the rest. This is often the case if the task is shared between the parties (e.g. the architectural and civil engineering designs are made by the transferee and he orders the technological designs from the supplier of the engineering). In such cases the transferor supplies a detailed engineering for the technology, but conceptional designs only for the rest, or yields the data necessary for it.

When the task sharing is more ramified, more care must be taken in the exact drafting of the agreement. When designing is task-shared, plan coordination meetings are unavoidable (*see also* paragraph 2.3 of chapter 2). In the agreement the place, the frequency, the theme and the cost-bearing party have to be fixed. The results of the plan coordination meetings have to be laid down in minutes which do not replace the agreements themselves. The plan coordination may however have a bearing on the agreement, but this is a more complex issue (the technical modifications may influence the price, the terms, etc.).

The language, copies, and delivery term of the documents are also to be fixed (*see* further details in paragraphs 1–1.5 of Chapter 2 and in paragraph 3.7 of this chapter) in connection with the liability for the result.

b. Technical assistance agreements

The technical assistance agreements have already been discussed (*see* paragraph 2.5 of chapter 2), as well as those clauses of other agreements (licence and know-how agreements) which relate to technical assistance. It is superfluous to discuss this matter here again. It has to be emphasised that while in case of such agreements all contingencies cannot be envisaged in the contracting phase, the more details are clarified (mainly in case of large volume services) the less problems there shall be at the site. The resolving of practical issues in the agreement is however not negligible for relatively minor services too.

c. Services of a commercial character

Such types of services are scarce in licence agreements, but often present in know-how agreements. The transferor may namely undertake to sell the goods produced under the licence or the know-how through his own sales organisation. The transferor sometimes performs a market research for transferee. It has also occurred that the transferor ceded a whole market to transferee, 'transferring' at the same time its representatives and customers as well. To services of a commercial character belongs that case also, where the transferor grants a licence for the use of its trade mark. There are contracts, where the main deal is the trade mark licence, and the technical matters are transferred in order to further the trade mark licence.

Commercial services are however different from case to case and the summarising of their contractual stipulations is hardly possible.

The same applies to agreements for the transfer of non-technical know-how, which relate mainly to commercial-like services.

d. Delivery of goods

In case of licence and know-how agreements it is not at all unusual, that the transferor supplies certain raw materials or parts to transferee, especially in the early phase of the realisation of the technical solution. In other cases he may deliver investment goods to the transferee. In respect of such collateral obligations the usual conditions of sale prevail in licence and know-how agreements. It has to be remarked, however, that the deliveries of raw materials or parts under the mentioned agreements may be limited (*see* paragraph 4.3 of chapter 3). The delivery of investment goods, though, often becomes a main obligation within the agreement while the licence or know-how is a collateral one.

e. The fulfilment of collateral obligations by third parties

It is important to survey the situation where the exclusive right – the patent or the know-how – is the property of one party, but another party is fulfilling the collateral obligations.

This case is often encountered if the licensor or the transferor of the know-how does not possess a design department, or the realisation requires a main contractor, but the transferor of the technology has no such organisation.

It is possible that in such cases two separate agreements are concluded, one between the licensor and the licensee or the transferor of the know-how and the transferee, the other for the designing work between the designer and the transferee. The machinery or other goods may then be delivered either by the transferor or by the designer – at the same time main contractor – but it may also happen that other companies are given this task. An advantage of this procedure could be, that it is sometimes cheaper, than purchase from a single source. Its drawback on the other hand is, that the liability is split when more partners are involved, which in case of unfavourable developments would make the liability issue a very thorny one indeed.

It may happen, that the transferor does not transfer the technical solution or the know-how directly to transferee, but to the designer and/or main contractor. In such cases it is the latter who concludes the deal with transferee for the designing work and the main contracting, and also for the transfer of the licence and/or know-how. The advantage of this solution lies in its undivided liability, its drawback in the higher cost.

A variety of this type of arrangement is one whereby, although the main agreement is concluded between the designer/main contractor and the transferee, and the licence fees or the countervalue of the know-how are/is also paid to transferor through him, the licence and/or know-how agreement itself is concluded between the transferor and the transferee directly. (The rights in connection with the licence and/or know-how are not transferred to the designer/main contractor, but to transferee, whereby transferee is directly responsible to transferor to maintain the secrecy, to not use the rights received beyond the agreed scope and to not enlarge the plant without the consent of transferor). Such contractual solutions are usual where the intermediary (the designer/main contractor) is a department of a machine manufacturer, who is not really interested in the use of a process technology.

This arrangement is advantageous for the transferee in several respects: the undivided liability is conserved, he has to make payments to a single address only (which can simplify matters in respect of credits or loans), and he is in direct connection with the transferor, which facilitates the transfer of improvements. Care must be taken regarding the price, and of course the direct agreement between transferor and transferee has to be drawn up very carefully to minimise the possibility of 'backing out' of the liability. Such licence and know-how agreements are often called 'by-pass' agreements.

Another variation on the same theme is, when the transferor and the transferee agree on the collateral obligations – services and supplies –, but the transferor does not enter into the agreement alone but in an *ad hoc* partnership with the designer/main contractor. This solution does not differ essentially from the one where the transferor plays the role of designer/main contractor too. The transferee has to take care that the members of the *ad hoc* partnership should be jointly and severally liable for the fulfilment of the agreement.

3.6.3. Other collateral obligations

The *transferability* of the licence, or the allowance to grant *sub-licences* – whether in the case of patents or of know-how – must be decided in the agreement. Should the contracting party not be identical with the end user (e.g. the contracting company is a commercial, the user is an industrial company) it has to be stipulated that the contracting company has the right to grant sub-licences, eventually only to certain users. There is no need to do so if the contracting company acts on behalf of the end user.

In case of patent licences it is important to stipulate which party is obliged to *maintain the patents* further and whether the licence may be *registered*. The registration is important for an exclusive licence, because the exclusiveness of the licence is valid towards a third party only if it is registered.

Similarly, in the case of patent licences the licensee will stipulate that he has the right to take steps against *infringers of the patent* – subject to legal provisions – should the patentee not take such steps. The registration of the licence is important for this contingency too, since many legislations grant this right only to the registered licensee.

3.7. Technical and other guarantees

With a patent licence or know-how in some cases a 'pig in the poke' has been bought, e.g. if the invention has not been adequately tested (laboratory experiments only were made and no industrial ones), the designs delivered are eventually useless, the projects cannot be realised in practice, etc. To ward off this risk the transferee will justifiably require certain guarantees in the agreement.

The transfer may be eased, if the transferor has a lot of references, proving the realisation of its techniques in a similar size plant and under conditions similar to the transferee's wishes. In this case an actual demonstration may be unnecessary.

3.7.1. In general about technical guarantees

The licensors and transferors will try to avoid or minimise the guarantees. In the case of patents it is sometimes simply said, that the transfer of the exclusive rights suffices.

The licensor's declaration that he has no industrial experience on the basis of which he could give various technical guarantees, is to be appreciated. In respect of know-how an *ad hoc* working committee of the United Nations[6] stated in its intermediate material that:

> 'The grantor on the other hand, (...) may have reservations regarding the grantee's competence; he may not be prepared to furnish the additional technical assistance that may prove necessary; (...) he may also consider that the nature of the know-how is such as to exclude any guarantee of the result, or he may have objections of principle.'

(The final version did not contain this text, although it is very true).

6. Economic Commission for Europe Industry and Materials Committee, Ad hoc Working Party on Contract Practices in Engineering, *Draft guide for the drawing up of contracts relating to the international transfer of know-how in the engineering industry*, an intermediate text of 5 December 1967 (manuscript).

The transferee, however, justly feels much safer if he receives guarantees, even against a somewhat higher licence fee or price. For this reason it is nowadays customary for the technical guarantees to be a substantial part of licence and know-how agreements.

Should the parties agree on the technical guarantees, the achieving of the parameters, as agreed upon, has to be proven. This may happen either in the plant of transferor or in the plant of transferee. If the demonstration takes place in transferor's plant the transferee must be convinced beyond doubt that the result demonstrated was due to the licence and/or know-how transferred to him. In the course of this demonstration the transferee may be assured that a working technical solution shall be transferred to him, but he cannot be entirely sure that the technical solution received can work equally well in his own plant. A much clearer situation is brought about if the demonstration is made in transferee's plant. If the licence and/or know-how is coupled to the delivery of equipment, there seems to be no other sensible choice.

3.7.2. Types of technical guarantees

The technical guarantees are related to requirements such as

a. quantity or capacity,
b. quality,
c. specific consumption and/or yield,
d. others (such as environmental requirements, the fraction of impurities, of by-products, etc.).

Ad a. The guarantee of *capacity* or *quantity* may be expressed in various ways. The actual formulation depends on the technical details. For a plant working continuously the yearly capacity may be determined, whilst in the case of batch production fixing the production per hour or per day is common. It must be taken into account also, when the plant, production line, production unit etc. is capable of continuous operation, that it may in practice be working only in one or two shifts, or that it may have to be stopped from time to time for maintenance purposes.

The kind of product is related to the capacity. This becomes important if more than one product is to be produced by the same plant. The capacity may however depend not only on the kind of product, but also on the raw materials. Should several raw materials be suitable with different capacities, different capacity guarantees should be set. The same applies to the auxiliary materials and to the variability of other conditions which may cause variations in the capacity. If there is no such variability, this circumstance shall probably be fixed as a precondition of the technical guarantee.

Ad b. The *quality characteristics* are very varied. Sometimes they can be expressed by a single figure, but more often the chemical and/or physical characteristics of the product have to be described accurately. Sometimes the maximum amount or the highest proportion of impurities will be characteristic, in other cases the quality of the product is determined by a set of physical parameters, etc.

Ad c. The *specific values* express on the one hand the quantity of the raw and auxiliary materials to be used, on the other hand the amounts of the different kinds of energy inputs necessary for unit production.

The raw material consumption, mainly in the case of expensive raw materials, is very determinative for the yield and the efficiency, hence the related guarantees are important. When the production is highly energy-consuming the minimisation of energy needs is of special importance.

The economic consequences of the use of materials or kinds of energy are not all of the same order, and those of minor importance may be neglected. In some technologies the use of one auxiliary material or kind of energy improves the efficiency of another one, and therefore it is not always feasible to determine the specific values per kind of material or energy separately. Formation of a so-called 'basket' is then usual. In this 'basket' the kinds of materials and energies do not appear with their respective quantities, but are included in a product reached by multiplying their quantity and their predetermined price, and the guarantee relates to the global value thus calculated (naturally with certain tolerances). The transferor has to demonstrate this value, but not its components.

Ad d. Sometimes *other* guarantees are also stipulated. Such are e.g. environmental ones, which nowadays become more important (e.g. the maximum impurity content of industrial sewage, or the pollution of air). In some cases there are also stipulations in respect of guarantees concerning auxiliary materials, which can become quite important in certain cases.

3.7.3. Demonstration of the guarantees at transferor

The transferor is sometimes willing to demonstrate only that the technology works in his own plants. This may be the case if an unpackaged transfer takes place. The transferor has no means here to control the activity of the transferee, or to influence it in any respect.

In such cases the transferor simply demonstrates that the technology adequately works in his own plant, and invites the experts of transferee into his plant, maybe together with some external experts, upon adequate secrecy agreements. Such demonstrations are however rather scarce, because there are few transferees who would be satisfied with such a demonstration.

3.7.4. Demonstration of the guarantees at transferee

The undertaking of the technical guarantees has no value without their demonstration. It is usual to demonstrate the guaranteed values at the transferee's plant. The reason for this is, that the operational conditions are generally not exactly the same there as at the transferor's site. In the case of complete plants there is no other possibility of demonstration than in the newly built establishment.

The demonstration of the guaranteed values takes place during a given stage of starting up, in the so-called guarantee tests. The start-up of the plant commences after completion of the erection of the equipment and it is a complex procedure, which may generally be divided into the following stages:

- cold tests (this process belongs in general to the erection stage, as its last phase),
- hot tests (i.e. tests where materials are also present),
- trial operations (this is really the commencement of putting into operation),
- steady operation and within it the guarantee tests, the demonstration of the guaranteed values (this is also called 'commissioning').

The expressions in agreements are however not uniform. Sometimes the whole procedure is called 'commissioning,' although strictly speaking only the last stage – the handing over and receiving stage – is the 'commissioning,' while the previous phases may be called 'start-up' or 'putting into operation.' To avoid such inaccuracies in an agreement, it is practical to explain the meaning of each expression used. This is also necessary because in most agreements the time limits are also coupled to the expressions. For example, it may be stipulated that the transferor has to prove the guaranteed values within 90 days from the start-up and has to 'commission' the plant within that period. It is hence important not only to define in an agreement what is to be understood under start-up, but also where it commences (e.g. do the cold and/or hot tests fall within this period or not). In other cases the time limit counts from the trial operation, or from the moment when the transferor declares that he is 'ready for commissioning.' This start-up period, like the period of the test runs, is short (e.g. 30 days) and depends also on how long a guarantee test run is required by the transferee. It may happen that instead of 30 days the mentioned period is only a continuous 8 days of operation, or only a 48-hour one, depending very much on the case. In any case the transferee has to be satisfied that the plant is suitable for continuous operation. Sometimes this may simply be proven by the production of a given minimum quantity.

The date of the commissioning (because this is the essential part), has not only to be bound to the commencement of the start-up, but usually also to the coming into force of the agreement. This is obvious when the transferee has practically no role in reaching the date of commissioning, or his input is time-limited. A typical example are the turnkey jobs, but such a time-limit stipulation may also be present in other cases (e.g. when the site is prepared for the reception of the equipment and there is little erection work to be done by transferee).

Normally it is not only stipulated within which time-limit the transferor is obliged to demonstrate the guaranteed values, but also which is the deadline after which he is not obliged to do so. For example, should the guarantee tests become impossible – outside the fault of transferor – after a certain period of time reckoned from the coming into force of the agreement, the transferor is freed from the demonstration obligation and the plant has to be regarded as commissioned. The motivation for this provision is that if the erection works are executed by somebody else than the transferor and are protracted to a large extent, the transferor should not be liable for it. A similar case arises if e.g. the raw material or energy is not available at the site for a long time and the transferor, through no fault of his own, is unable to demonstrate for that reason. It is unfair for the transferor to be under obligation for an unlimited time.

It is important that the start and end points of the demonstration procedure should be defined exactly. It must also be fixed in the agreement what happens in case of an unsuccessful test (i.e. if the transferor is unable to demonstrate the values): does he have the possibility of repetition (an extended time limit) and if so, for how long and how many times may he repeat the demonstration. At least one repetition possibility is normally given to every transferor.

In practice a number of guarantee test runs fail due to the fault of transferee or a common fault of transferor and transferee. It may e.g. happen that during the 30 days of test run the steam supply stops (if the boiler plant was not delivered by the transferor) and hence the guarantee test has to be stopped. It depends on the kind of the technology whether the test may be continued when steam becomes available or has to be started again.

Such situations must be considered when contracting and it should be fixed what happens if the test run is to be suspended, both if this is transferor's fault and if it is transferee's fault.

Depending on the technology, the test run period may also be regulated contractually so that its validity is coupled to a certain number of hours. Should e.g. a 30-day continuous test run be discontinued by the fault of the transferor, but should the stops be not more than x hours each, or in total y hours (or days), this will not invalidate the test, while if the stoppages surpass this time limit, the test should either be prolonged by the time dropped out or is to be regarded as unsuccessful.

Should the trial test be discontinued because of transferee's fault, the transferor is guiltless, but he may be obliged to at least a 15-day trial run in that case. Naturally any other logical solution is feasible, depending on the case.

Not only the duration of the tests is to be fixed – it may well be that the requirement is not a single 30-day test, but several shorter series of tests, with various raw materials and towards various end products – but also the methods of measuring the guaranteed values and the tolerances of the measurement.

In respect of the way of measuring it has to be clarified, whether it takes place continuously or whether samples are taken from time to time. The measurement of quantity is usually continuous, while that of quality may be intermittent. Should the control be a time-to-time one, the intervals of the measurement have to be determined. Often the methods of measurement must also be clarified, since it may happen that one obtains different results with different methods of measurement.

Tolerance has a dual sense. One may speak of a tolerance of the measurement (due to the inaccuracy of the methods of measurement and the instruments) and a tolerance of the values themselves (the allowed deviations from the values to be demonstrated). It may happen that both are given together as a lump tolerance sum.

The whole demonstration period ends with the commissioning, when the demonstration of the guarantee test has been successful. It is obviously successful, if all the guaranteed values measured by the previously agreed methods are within the limits allowed in the agreement. In this case minutes have to be drawn up on the commissioning.

Should any of the guaranteed values lie outside its contractual limits, the guarantee test is to be regarded as unsuccessful and the agreement has to provide what should happen (the general solution is repairing, rechecking and a repetition of the guarantee test within an agreed time limit). It may happen, however, that only a non-important guaranteed value is not reached and this only by a little. It is then economical to not repeat the whole guarantee test for such minor reasons, but to commission the plant and to request either a price reduction from transferor to compensate for the defect, or to allow the transferor to eliminate the deficiency by making the necessary smaller adjustments at a suitable later time (e.g. when the plant is stopped for maintenance purposes anyhow). Such fine points need not be taken care at contracting, but may be settled amicably in a later agreement. (The consequences of unsuccessful demonstration were already discussed in paragraph 6.5 of Chapter 3).

3.7.5. Preconditions of demonstration

A stumbling block may arise when the transferor requires for the demonstration certain preparations, such as that:

- the raw material, auxiliaries and the kinds of energy as specified in the agreement are at his disposal,
- the trained, skilled personnel and the unskilled labour is available,
- if the equipment or a part of it are not delivered by transferor, they must be in conformity with the specifications of the transferor and he has to be convinced thereof,
- in general, all the conditions are present for a continuous operation.

Besides, the transferor will often insist that the whole process of putting into operation and the demonstration has to be done under the technical direction of his experts.

These conditions are to be regulated as exactly as possible in the agreement. It may happen however that all the conditions cannot be set down at contracting or that some of them demand an intimate know-how which the transferor is ready to disclose only after the coming into force of the agreement. In such cases it is important to fix at least when these conditions shall be communicated, although there is naturally no possibility to postpone the stipulation of basic conditions (e.g. it cannot be communicated later than at contracting what type of raw material has to be used).

It may happen that the transferee is unable or only partially able to secure the necessary conditions laid down in the agreement. It is obvious that if the basic conditions are not fulfilled, the transferor has no obligation to undertake the demonstration. Moreover he must refuse it, if the lack of the adequate preparation will probably lead to a fiasco. However, he will use this possibility only as a last resort, when there is no temporising solution available, for fear that only the fact of non-demonstration shall become known in business circles and not its reasons. Such a temporising alternative may be that in the absence of the preconditions, transferor does not undertake to demonstrate the values fixed in the agreement, but lower ones. For example, if the raw material fixed in the agreement is not available, he may undertake the demonstration of a worse result.

When the raw materials are minerals, such situations may so often occur that the parties will consider them at contracting. Thus the guarantees may be given for a certain kind of raw material, but it will also be laid down that when only a different raw material is available, they should agree later on corrected guaranteed values. It is possible to assign different critical values to the different kinds of raw materials in the original contract too.

If however the lack of the preconditions unequivocally excludes a successful demonstration, the transferor will surely not perform one and due to the general conditions mentioned above he will be exempted from the demonstration of the guaranteed values. Nevertheless in such cases the transferor may be willing to undertake, if the preconditions are later met, to put the plant into operation within the frame of technical assistance, including tests, but of course without legal consequences.

3.7.6. Securities for the demonstration of guarantees

The stipulation in the agreement is not in itself a security for the fulfilment of the technical guarantees. This is the reason why the transferee will try to obtain various securities. The simplest guarantee is, that the transferee makes payment of a portion of the licence fee or of the countervalue of know-how dependent on the fulfilment of the technical guarantees. Should the licence fee or the countervalue of the know-how be a fixed amount, it is usual to assign a fraction of the payment to the performing of the training and another fraction to the fulfilment of the guaranteed values. Should the payment be a function of production (be a royalty), an advance payment is to be paid in such cases too, but the transferee often stipulates that payment is made only upon the demonstration of the guaranteed values (except if the non-performance is his failure). Should the transfer of the licence or the know-how be coupled to the delivery of equipment or of a complete plant, it is customary that the parties agree on certain securities bound to the fulfilment of the guaranteed values (e.g. a bank guarantee) and a fraction (say 10–15 per cent) of the price shall become due only after the fulfilment of the technical guarantees.

The parties may agree on special conditions too. It may happen that the transferee settles his payments not in cash, but with the products produced by using the licence and/or know-how according to the agreement. In case the transferor guarantees that one main characteristic of the product produced with the use of the licence and/or know-how shall be a better quality than those of the competition, this condition will be very favourable for the transferee, because it is in itself a certain security that the transferor will assist to attain this quality, since he will otherwise miss payment of the full licence fee and/or the countervalue of the know-how.

Another contractual possibility is that the transferor does not transfer its technical solution (a patent or know-how) against cash, but founds a new company together with transferor, the equity capital being supplied by the transferee but a part of the shares being given to the transferor. It is obvious that here it is in the interest of the transferor that the technical guarantees should be reached in the new plant; moreover that they be reached soon.

3.7.7. Non-demonstration of guarantees

In agreements where technical guarantees are stipulated, the consequences of non-fulfilment (the sanctions) must be regulated also in a suitable way.

The situation is namely quite different if the deviation from the guaranteed values is only a minor one than when it is a major one which may even make the technical solution unusable. While in the first case penalties and/or liquidated damages (their amount is to be fixed in the agreement) are stipulated, in the latter case as a rule the payment of damages is required while the transferee is given the right to terminate or to repudiate the agreement.

There are agreements where the parties do not determine the consequences beforehand but stipulate that in such cases an arbitration court will decide on the further destiny of the agreement and, depending on the case, in all other respects. (We have already discussed other aspects of this issue in connection with the sanctions for negligence in paragraph 6 of Chapter 3).

3.7.8. Other guarantees and warranties

In licence and know-how agreements it may happen that the transferor not only undertakes technical guarantees, but *economic* or *marketing guarantees* too. Such guarantees are very risky and transferors will undertake them only exceptionally.

Should the delivery of machinery also take place by the transferor, there will also be a so-called mechanical guarantee. (This shall be discussed in detail in paragraph 4.1 of this chapter.) Over and above that the transferor of the technology has to warrant, that the transferred technology does not infringe third party rights. Paragraph (1) of § 18 of the Hungarian Patent Act (and paragraph (1) of § 28 of the new Patent Act of 1995, already mentioned) prescribes the following warranty of title:

> 'The patentee warrants during the whole duration of the agreement, that third parties do not have any such rights in connection with the patent, which would hinder or limit the use of it.'

Should there be such rights, the licensee may terminate the agreement immediately. This stipulation is eased by § 22 of the Patent Act (and paragraph (1) of § 32 of the new Patent Act of 1995), according to which when a patent is not usable without infringing the rights of others, a compulsory licence has to be granted for that hindering patent. (We have already discussed compulsory licences in paragraph 12.1 of Chapter 1.)

Finally paragraph (3) of § 18 of the Patent Act (and paragraph (1) of § 28 of the new Patent Act of 1995) prescribes that the patentee is obliged to inform the licensee about 'eventual rights and important circumstances.'

Due to the fact, that the transfer of technology is in most cases international, this warranty deserves a special interest. Should the subject of the transfer be either a patent or know-how, this may infringe third party rights in the transferee's country. A patent research in this respect is unavoidable, first of all in the transferee's country and, should export be envisaged, in all the concerned countries. It is another issue, that a compulsory licence will be granted in the majority of countries only for patents which hinder the utilisation of another patent.

As thoroughly as the transferor may investigate the hindering circumstances in a given country and find that there are none, there may still be some. In such cases the injured party shall claim against the transferee. Therefore the agreements must stipulate that the transferee upon receiving notice of a possible infringement is obliged to communicate this fact immediately to the transferor. In such cases the transferor shall be the person undertaking the necessary steps, perhaps involving the transferee too.

3.8. Other conditions

In connection with licence and know-how agreements we have to mention also some general, mainly juridical, conditions. We have to discuss first of all the breach of contract, its consequences, and the exemption from the obligations of a contract.

It is preferable to stipulate in every licence and know-how agreement the consequences of a *breach of contract.* The transferors' breach of contract arise from a simple delay, the failure to transfer stipulated information (a document) or a partial failure to do so, the failure to demonstrate the technical guarantees (*see* paragraph 3.7.7 of this chapter), the failure to fulfil the warranties, etc. The gravest breach of contract on the side of transferee is the non-fulfilment of its payment obligation. Another grave breach may be in know-how agreements the violation of the secrecy obligation. In all cases a breach occurs when the limitations prescribed in the agreement are not lived up to, provided such limitations are validly stipulated.

The consequences of the breach of contract on transferor's side are in minor cases a penalty or liquidated damages, in grave cases the payment of damages and the termination of the agreement or its repudiation. Breach of contract on transferee's side may involve – depending on its graveness – the payment of damages or the termination of the agreement by the transferor.

The parties normally agree that *force majeure* (Act of God) is an exemption cause, and in such cases they give a certain period of grace to each other for the fulfilment of his obligations, but they terminate the agreement after the expiry of this period of grace. In that case there is no obligation to pay damages.

The *approval of the authorities* (permit) belongs to the general stipulations as a precondition for the coming into force of the agreement. An approval is still necessary in a lot of countries, first of all in respect of licence agreements (though no longer in Hungary, where only a posterior notice is to be given for statistical purposes). Should such approval be necessary, the agreement does not come into force with its signature but only after official approval. The conditions of the coming into force and the consequences of the authorities' refusal have to be clarified in the agreement itself.

It is also important to clarify the *duration* of the agreement. The final date of a licence agreement is usually the validity date of the patent, but it may also be concluded for a shorter time. In the case of a know-how agreement it is not so simple. The transfer of know-how may be a one-time transfer or it may be a regular service, lasting for many years. Moreover certain rights and obligations may continue to exist after the expiry of the agreement. In know-how agreements this is so for the obligation of secrecy and the right of the continuous use of the know-how, in case of a licence agreement – should the licence expire prior to the expiry of the patent – the right of further use or the use for at least those products whose production is in course. These rights and obligations should be fairly regulated in the agreement.

The agreements normally end with *miscellaneous* and *closing stipulations*. Taxes and duties are among them (these are high in some countries, so it is important who should bear them), as well as the modification and transfer possibility of the agreement. Such miscellaneous and closing stipulations also include the language of the agreement, the correspondence addresses of the parties, the enumeration of the appendices, the date of the signature, the number of copies signed and finally the signature.

These conditions are to be found in any agreement and are not typical of licence and know-how agreements, and we shall not discuss them in detail. We shall however discuss (in paragraph 4.2 of this chapter) the conditions of payment.

4. LICENCE AND KNOW-HOW AS COLLATERAL OBLIGATIONS

If a patent for which a licence is granted, or the know-how to be transferred, is additional to an agreement for services or an agreement for the delivery of investment goods, it appears in the agreements as a collateral obligation only. This is characteristic of engineering agreements or agreements for the delivery of various equipment (generally a complete line of machines, or a complete plant). The know-how is often included in the documentations to be delivered and in the training and does not even have a separate price tag.

Sometimes it is mentioned that the price contains the know-how as well, but no sum is indicated. The know-how becomes in such cases an inseparable part of the service or delivery of goods making up the main subject. It is impossible to determine, out of the documentation to be delivered or the training to be effected, what portion belongs to the category of know-how. This circumstance may simplify the issue of the technical guarantees, but at the same time it may be disadvantageous both for transferor and transferee.

On transferor's side it is a drawback that there is no protection of the intellectual property transferred (there is e.g. no secrecy obligation), and it is even not secured that the transferor would receive, in case of the enlargement of the plant, any money. This seems to be advantageous for the transferee, but the drawback for him is that should there be an obligation to pay customs duties for machinery in international trade, he has to pay them even on the intellectual property, because not having a separate price it is declared practically as machinery. Further drawbacks for the transferee are that he does not receive any improvements and that the transferor may sell without any restrictions the same plant to the competitors (together with the same intellectual property). Summing up: the rights and duties of the parties in respect of the intellectual property are not clarified.

In some contracts where the patent licence is a collateral obligation in an agreement on engineering or delivery of equipment, it is mentioned when stating the price that the licence fees are included, but it is nowhere mentioned to which patent(s) it relates and of which patent(s) the licence fees are included in the price, nor are the duties and obligations of the parties clarified in connection with the patent(s). Such vague 'licences' cannot be registered.

It may be acceptable that in an agreement the existence of know-how as a collateral object is omitted, but it is not acceptable that licence rights to patents which are valid in the countries where the equipment to be delivered shall operate, are mentioned only in hazy outlines. It is much safer if at least a chapter of the agreement deals with the patents (e.g. the chapter dealing in general with intellectual property). The agreement should contain at least the registration number of the patent, the fact that transferor has granted a licence for it, and what the character of this licence is. If there are any limitations, these must also be enumerated and it must be clarified who is obliged to maintain and to defend the patent(s). Should the global price include the licence as well, this need not be regulated separately, nor need the technical guarantees which are derived from the equipment purchased.

The issue of warranties, too, have to be clarified in the intellectual property chapter (including the validity and the title). The general juridical conditions are normally contained in the other parts of the engineering and/or delivery of equipment agreement.

If the collateral obligation is transfer of know-how and it is important to regulate the issues in connection therewith in the agreement, the following procedure may be advisable. The documentation should be arranged into two parts, the first concerning the engineering and the delivery of machinery only, the second having a know-how character (e.g. technical descriptions). This arrangement may be expressed in the payment conditions too, e.g. a certain payment may be coupled to those parts which have a know-how character. The training itself in the plants of the transferor may be limited so that only so much knowledge is transferred as the transferor deems absolutely necessary.

A secrecy clause may be present in the agreement. (The existence of such a clause implies that know-how is also a subject of the agreement). Should this be necessary, limitations of the amount to be produced or territorial limitations may also be stipulated.

A stipulation according to which the buyer has no right to enlarge his plant is not necessarily a consequence of the existence of know-how; it may well be a limitation forbidding the repeated use of the design. Should the agreement mention however that the establishment of a new plant according to the technology transferred is subject to a separate permit of transferor (and requires an extra payment), that implies that know-how has been transferred under the agreement. By contrast, if it is laid down in the agreement that the enlargement of the plant is not allowed even by applying other technologies (whether that of the transferee or of a third party), this does not allude to the protection of intellectual property but is a purely commercial limitation (a limitation of competition, which may be against the antitrust regulations in many countries).

The above agreements, in which the patent licence or the transfer of know-how are collateral obligations, are not licence or know-how agreements *per se.* They are basically contracts of sale or entrepreneurial ones, the main subject of which is engineering and/or the delivery of goods (equipment). The patent licence or the transfer of the know-how is merely additional. Such typical agreements are those which relate to the *delivery of a complete plant.* Such deals are connected – if not directly – to a patent licence or to the transfer of know-how, but it is of lesser importance. Any kind of know-how (*see* paragraph 5 of Chapter 1) may be involved in such cases. Therefore it is understandable that in such contracts the condition is always included that the enlargement of the plant delivered according to the transferred design and technology, and the establishment of a new plant on the same basis, are bound – at least within a certain time limit – to the permission of the transferor and to a repeated payment (although this is normally less than the one for the original licence).

The structure of agreements for the delivery of complete plants is generally the following:

- the denomination of the parties to the contract,
- the subject and aim of the agreement,
- the detailed description of the engineering activity,
- stipulations in connection with the intellectual property (e.g. patents),
- the detailed regulation of the delivery of goods (including parity and transport issues),
- the detailed regulation of services (training, direction and/or supervision of the erection and the putting into operation),
- technical guarantees,
- so-called mechanical guarantees,
- sanctions,
- the price,
- payment conditions,
- bank guarantees,
- general juridical issues (warranty of title, *force majeure*, the governing law, arbitration court),
- the coming into force,
- so-called miscellaneous and closing stipulations, and
- the enumeration of the appendices.

These are the issues which must be decided unambiguously. The mentioned sequence is derived from the practice, but any other logical structure is feasible. The sanctions may e.g. not form a single summarising chapter, but may be discussed in the relevant chapters (e.g. the delay of the delivery of documentation in the delivery of documentation chapter, the delay of delivery of goods in that chapter, the non-fulfilment of the technical guarantees in the guarantee chapter). Sometimes the contractual price is established at the head of the agreement, or the issue of the intellectual property is put among the miscellanea because it is of minor importance in the deal. The sequence does not determine whether the agreement is a good or a bad one. Sometimes the contract is so arranged that some of the subjects are not treated in the main agreement, but in separate ones (e.g. agreements on services or intellectual property are to be signed simultaneously with the main agreement).

Most of the individual subjects have been discussed already, and in the following we shall deal only with two of them: the mechanical guarantees and the price and payment conditions.

4.1. Mechanical guarantees

A mechanical guarantee is an obligation to eliminate failures (the reparation) of the equipment delivered.

It is obvious that the starting time, its expiry, and its duration have to be fixed in the agreement. The duration is in most cases 12 months from the start-up of the operation (*see* paragraph 3.7.4 of this chapter) and it has to be clearly stated whether this is counted from the first test or from the commissioning.

It is usual to limit the mechanical guarantees. It is often stipulated that their duration is 12 months from a given date, but not more than 24 months from the last delivery of equipment. This means that should the 12 months expire prior to the mentioned 24, the guarantee then expires, but if the 24 months would expire earlier, the latter shall prevail and the effective guarantee period therefore becomes shorter than the mentioned 12 months. This may occur if the erection period is very long.

Depending on the difficulty and the complexity of the technical work 24 months sometimes is too short a time for performing the technical work, the commissioning and the monitoring for some time of the faultless operation of the plant. A longer period, usually 30 months, may then be agreed.

Care must be taken if the delivery of the plant and its putting into operation takes place in two stages. A single guarantee period is irrational in such cases.

More than one expiry date is also contained in agreements where besides the delivery of the plant the delivery of certain auxiliary equipment also takes place. The mechanical guarantee dates of the latter are not equal to those of the whole plant, but are determined separately depending on their delivery and commissioning.

The expiry of the mechanical guarantee may also depend on the failure of the guarantee tests (the commissioning). Stipulations are also not rare which provide that in case of an unsuccessful guarantee test the guarantee period (the mentioned 24 months) is to be prolonged.

The mechanical guarantee extends normally to the quality of the material, to the construction, and the quality of the goods produced. (In respect of the latter the specifications and standards will prevail and this is why these are important). This means that in case of a failure of the equipment, the seller (transferor) is obliged to eliminate the failure (by reparation or replacement). The mechanical guarantee however does not extend to all failures and excludes those which are the consequence of unwarranted intervention, non-adherence to the operation and maintenance prescriptions, natural disasters (flood, earthquake, lightning) or inadequate storage. The guarantee also does not extend to those parts which are subject to tear and wear (e.g. wearing of a bearing ball, natural wear).

It is up to the seller (transferor) to decide whether a reparation or a replacement should take place. The seller usually sends an expert for reparation – free of charge – and in case of replacement he undertakes the transport costs of both the new piece and of that to be replaced. It is also a usual condition, that the duration of the mechanical guarantee shall be extended by the duration of reparation and in case of a replacement it commences again.

The buyer's (transferee's) obligation in connection with a failure is that he should report it immediately, should enable the inspection of the failure at site by the expert of the transferor, and eventually determine its cause and undertake the necessary steps.

It is usual to give a mechanical guarantee for used equipment too, but this is usually shorter (e.g. 3 or 6 months).

4.2. Price and conditions of payment

In most agreements the price is set as a single global sum. This sum may however be divided into components and if more than one service and delivery are dealt with in the agreement, this is the recommended solution. Out of a single global sum it cannot be determined whether or not the engineering costs or the licence fees are included.

The following services and deliveries may be included in the global price, depending on the circumstances:

- a patent licence fee or the countervalue of the know-how,
- the engineering, and within this the price of each stage (basic, detailed engineering),
- the training,
- the direction of erection,
- the direction of putting into operation,
- the designers' supervision,
- other services,
- the price of the delivery of goods (separated e.g. into equipment, auxiliaries, spare parts, etc.),
- the countervalue of possible leases (e.g. the lease of the erection machinery),
- the countervalue of other activities.

It may happen that the global price does not contain the precise countervalue of any of the above-mentioned items, but a frame amount only. This is usual especially for the services expressed in man-months. Such frame amounts may also be determined e.g. for spare parts, for the training, etc. It is also customary to put into the price a budgeting amount for unforeseeable items (like e.g. the uncertain length of a water pipe line).

It is very important to define the price accurately and also to fix in which instalments it should be paid and when it becomes due. It is also a possibility that the price is expressed in different currencies, e.g. the daily allowances of the experts may be expressed in local money and the rest in US dollars. It is also important to determine whether the given price is a fixed amount or a gliding one, and if the latter, how.

Should the agreement concern a transfer lasting for a substantial period of time, it is unavoidable that the original price should be modified due to market price changes, details of which are to be fixed in the agreement.

The *conditions of payment* are quite intricate. We shall deal here only with the payment in cash, the treatment of credit issues going beyond the scope of this book.

4.2.1. Partial payments

The payment in cash may consist essentially out of three parts, which are however staggered in time:

a. the advance payment,
b. the instalment to be paid at delivery (fulfilment), and
c. an amount withheld (retention).

Ad a. The advance payment
The usual measure of the advance payment is 5 to 20 per cent. Its payment is normally bound to the coming into force of the agreement. When the advance payment is on the high side, it is customary to pay it in two or three stages, e.g. 10 per cent is to be paid after the coming into force of the agreement and 5 per cent x months thereafter. This is similar to the arrangement preferred for long delivery periods, in which case before the delivery so-called *progress payments* are made, i.e. advance payments, to enable the production of the equipment etc. to be transferred. This payment may be bound simply to the lapse of time (x months after the coming into force of the agreement a certain amount is due and y months later another amount), but the payment of such instalments may also be bound to progress reports which certify the progress both in the case of an engineering and of a machine manufacturing activity.

Ad b. Partial payment at delivery
The main part of the price (countervalue), usually at least 80 per cent, is to be paid at delivery or partial delivery. The transferor usually asks for some securities for this payment such as letters of credit (L/C, whether confirmed or unconfirmed), a commercial letter of credit, a bill of exchange (B/E), a promissory note (in most cases a guaranteed one), or a bank guarantee.

Even if securities are postulated in the deal, the agreement has to state clearly in what cases the transferor shall get his money, because the remittance is made in most cases through a bank. The bank deals only with documents and neither knows what the documents relate to nor sees the goods; it cannot judge when the security becomes due.

In case of partial deliveries it also has to be stated clearly, whether a payment becomes due and in which fraction of the total amount. If the delivery of documentation is part of the deal, and if this is to be paid separately, this must be fixed also.

Since securities have a validity period, one must take care to prolong them in case of a delay of delivery (especially if the delay is not the transferor's fault). Otherwise the transferor will have to deliver without security (so-called 'open deliveries'), which is risky.

Ad c. Retention

The transferee tries to secure for himself the goods and services in accordance with the prescriptions of the agreement in due time. To that end – *inter alia* – he may retain a fraction of the payment (countervalue). Should the commissioning be frustrated by transferor's fault, the latter will thus not be paid fully. The retention is normally 10 per cent but it may be more; in a few cases it is less. The payment is bound to the minutes of the commissioning.

The transferor will attempt to secure that when commissioning is delayed not by his fault, he will receive the amount retained at the date originally foreseen. This is the reason why the period of retention is limited. The bank holding the money does not investigate the deal *in merito* (who should be held responsible for the delay) and therefore a practice has come into use, whereby the retained amount may be paid either against presentation of the minutes of commissioning, or after the elapsing of a certain period (depending on the case this may be 6 to 12 months after the last delivery) even without the presentation of such minutes. This payment however does not take place if the transferee testifies – e.g. by presenting a copy of a letter written to the transferor – that the commissioning has been and is still delayed by transferor's fault.

4.2.2. The structure of the payment conditions in case of cash payment

The payment conditions clause of the agreements starts always with a statement about the currency in which the payments are to be made. In the majority of cases this is the currency in which the contractual price has been determined, but not necessarily. There are contracts where a part of the contractual price is to be paid in another currency (e.g. to a sub-contractor of the transferor) or where payment in more than one currencies was agreed at the outset. (This may well be the case if more than one country participates and credits or loans are used for the deal).

The payment has to be regulated here for each of the different currencies in the agreement. It is a matter of taste whether under each currency the corresponding partial payments are listed or *vice versa*, i.e. first the issue of the advance payment is regulated per currency, thereafter the payments at delivery per currency, etc. In practice the person who draws up the agreement is influenced to a certain extent by the question whether the same kind of services and/or deliveries are to be paid in several currencies, in which case he will list the currencies per kind of payment, or whether a single item of service and/or delivery is to be paid in a certain currency while another currency is attached to other services and/or deliveries, in which case he will prefer to group the payments by currencies.

For the sake of simplicity we shall hereunder only deal with payment in a single currency.

Advances are paid upon an invoice and a guarantee for the advance payment, to be given by transferor to transferee. In this guarantee, a bank or another financial institution – maybe the parent company of the transferor – guarantees that transferee shall get back the advance payment if the agreement is terminated through the transferor's fault, or if the delivery does not take place at the ultimate date, etc. The value of this guarantee normally depreciates proportionally to the effected deliveries.

When drafting the text of the agreement the following have to fixed in connection with the advance payment:

- at what time is it to be paid (at the coming into force of the agreement or e.g. within 30 days thereafter),
- in the case of payment in instalments, when these become due,
- which are the necessary documents.

Should the transferee not pay the advance in due time (and should this not be a precondition for the coming into force of the agreement), in the agreement there will be a stipulation that, should the payment of the advance be delayed, the delivery dates shall be postponed accordingly, and should it not be paid before a certain date, the transferor has the right to cancel or terminate the agreement. To avoid such complications it is to be advised, that the advance payment should be made a condition for the coming into force.

The *payment at delivery* depends on what type and how many deliveries shall take place: whether documentation is delivered, in case of engineering, whether it has a separate price tag, etc.

Should the engineering or the delivery of documentation have a separate price tag, the amount to be paid at delivery has to be broken down according to the items.

The maturity date, the way of payment, and the documents to be presented have to be clearly indicated. The way of payment is generally the same for each item (an L/C or a commercial L/C). Often the payment per item is made with the same L/C or commercial L/C. Therefore in both the agreement and the bank instructions the countervalue of each item is to be expressed in total figures, so that the banking institution shall be able to control the payment in formal respects. The payment instructions are characteristic to the deliveries effected, so different documents should prove the delivery of documentation and the delivery of goods, etc.; in other words, characteristic documents are to be assigned to the various kinds of services.

It may sometimes also happen that the countervalue of the documentation is not fixed separately. Should however the transferee want to be assured that the transferor gets his money only upon delivery of the whole documentation, he should include a stipulation in the conditions of payment according to which the transferor must declare that he has delivered the whole documentation. This declaration has to be presented, when partial deliveries are agreed, with the last delivery. Conversely, the countervalue of one or more partial delivery may also be bound to the declaration of the effected delivery of some partial documentation.

The concept of partial delivery being very vague in certain cases, it has to be clarified in the agreement. A way of doing so is by determining that only the delivery of complete units of equipment (so-called complete posts) may be regarded as partial deliveries entailing payment.

In connection with engineering activity it also happens that the payment of its countervalue is not made synchronously with the various partial deliveries, but simply because of the lapse of time, e.g. in monthly equal rates. However, it is in such cases a usance to bind at least certain payments to certain partial performances, e.g. the payment of a certain sum to the delivery of the basic engineering, etc.

As regards the *retention*, it is usual to stipulate *in lieu* thereof a bank guarantee (a so-called performance guarantee) given by the transferor. The giving of this guarantee can be brought in conjunction with the effected payment of a certain amount by transferee. Due to the fact that this guarantee costs money to the transferor, the more in case of an extended delivery, transferors prefer to give such performance guarantees not at the coming into force of the agreement, but at a later date only. Such a possible later date may be that of the first use of the L/C, or after the using up of a certain amount from the L/C.

There are also cases where a retention was originally stipulated, but because of a delay in the commissioning (not by transferor's fault), parties later agree that the transferor gets the retained amount against a performance guarantee to the same amount as the retention. Such cases are however rare.

It may also happen that the transferor receives the amount of retention upon presentation of the minutes of commissioning, not unconditionally but only against a bank guarantee which amounts to a fraction of the retention amount and which shall be valid up to the expiry of the mechanical guarantee. In such cases a bank guarantee of this type has to figure amongst the documents for payment of the retention. It is to be remarked that the performance guarantee may be a B/E instead of a bank guarantee.

Two stipulations normally appear among the payment conditions: one laying down that the transferee has the right to deduct from his payments any *penalties* which are due, the other disposing about the *banking charges.*

Should the agreement also extend to the delegation of experts (in connection with the erection or otherwise), and should this have a separate contractual price, the conditions of payment therefor must also be regulated (e.g. it is paid per man-month). The agreement stipulates in such cases only that the payment becomes due within 30 days from presentation of the relevant invoices. This is in essence a payment against documents even if the payment would take place from an L/C. The documents to be presented have to be clearly listed. It may be feasible however that in the L/C itself a frame amount covers the delegation of experts, in which case the payment is to be effected in this sense.

5. THE FRANCHISE AGREEMENT

The franchise agreement relates essentially to the transfer of non-technical know-how combined with the licence of a trade mark. Dr. A. Vida writes:

> 'There is a close relationship between the franchise and the trade mark licence agreements, the franchise agreement being however more complex than the trade mark licence ... In a franchise agreement ... the grantor of the franchise transfers above the licence for a trade mark also his methods of management, experience in publicity, market research and other services to the transferee, who pays for them with a part of his business profit.'[7]

In the case of a trade mark licence the transferor may not interfere in the operation of the business of the transferee, except that he will probably reserve his rights to control the quality of the product. In case of a franchise agreement the service rendered by the transferor is accurately defined: the transferor of the franchise prescribes the franchisee's activities in detail, the transferee introducing any new ideas only within this frame. The transferor strictly controls the quality of the service of transferee.

A typical example of such franchise agreements is an agreement of the use of a name for a hotel. The main advantage for transferee is that he becomes the member of a hotel chain while his own identity can be emphasised too by putting a distinctive sign after or before the name of the chain (e.g. Atrium Hyatt). The guest who uses the service of the hotel may be sure (and also expects) to be serviced like in any other hotel of the Hyatt chain.

7. Dr. A. Vida, *Das Warenzeichen in der Wirtschaft der sozialistischen Länder*, C. Heymans, Köln, Berlin, Bonn, München, 1992, p. 142–149 (quote translated from the Hungarian version of the book).

The transferor of the franchise grants the use of its trade mark or trade name, but the main point is that it is transferring a diverse technical and non-technical know-how, starting with the experience of building a hotel up to its operating. This commercial know-how is another characteristic element of franchise agreements.

The transferee of the franchise is obliged to follow the style of the franchising chain and to operate his hotel according to the agreement. This obligation is offset by the advantage of getting guests through the chain. Stipulations are also usual whereby the transferee shall abstain from competing with the other members of the same chain and shall respect the recommended prices. He has also to participate in the common costs of the hotel chain in question.

Franchise agreements are typical not only for the hotel business but for the service industry at large (e.g. petrol stations, car-leasing companies, electronic services, restaurant chains such as McDonald's and Burger King). In some service branches a daughter-parent company relationship is more usual, although here too franchising may take place between the trade mark licensor and licensee. Agreements whose aim is the founding of companies (joint ventures in most cases) shall be discussed in Chapter 5.

Due to the fact that franchise deals are very specific and are often concluded within a packet of deals (i.e. besides the general agreement, a separate trade mark licence agreement and various separate agreements for the transfer of technical and non-technical know-how may be concluded), we do not enter into a more detailed discussion of the different kinds of franchise agreements.

5 Transfer Through Founding of Companies

1. JOINT VENTURES

The transfer of technology is as current between domestic companies in Hungary as it is in many other countries of the world. As there is however a tendency to found joint ventures between Hungarian and foreign companies, we will discuss first of all this subject, which could be generalised for other countries in a similar situation too.

There are various provisions of law dealing with economic associations all over the world. These are contained e.g. in Swiss substantive law (*Obligationenrecht*) and in Germany in commercial law (*Handelsrecht*).

The reason for discussing the Hungarian Act VI of 1988 on Economic Associations, as well as Act XXIV of 1988 on Foreign Investments in Hungary, is that both were purposely enacted to promote joint venture associations in Hungary.

According to paragraph (1) of § 1 of Act XXIV, an 'investment of foreign investors in Hungary shall enjoy full protection and security.' There are three possibilities for foreign participation:

- the foundation of a Hungarian company with foreign participation,
- the formation of a Hungarian company with 100 per cent foreign shares,
- a foreign participation in an (existing) Hungarian company.

There is no need for a permit for participations up to 50 per cent, only if the share is larger or a fully foreign company is to be founded in Hungary. The third mentioned possibility means that foreigners may buy shares in Hungarian companies. They may stay in the background in respect of their property, but they may also come to the fore and take part in the management of the company.

All these forms are suitable for transmitting new technology, know-how, and business methods to domestic companies, to facilitate their economic development in this way.

Joint ventures may also get considerable tax reductions, especially in their first five years. They may import goods needed for investment purposes, under appropriate conditions, duty-free and may repatriate their profits. Joint ventures have the full guarantee of the State and have the right to take their profits and transmit money abroad in foreign currencies (in the same currency in which their capital was brought in).

From the point of view of technology transfers it is of minor importance, that companies may also be formed in customs-free zones; these are to be regarded as extraterritorial companies.

1.1. Economic association forms existing in Hungary

There is nothing special about the Hungarian Act VI of 1988 (with legal validity from 1 January 1992), because similar forms of economic associations are well known all over the world. The aforementioned Swiss and German legislations also know such forms. The following economic associations may be founded in Hungary:

a. an unlimited partnership,
b. a deposit (or limited) partnership,
c. a union,
d. a joint enterprise,
e. a limited liability company (LLC),
f. a company limited by shares (Co./Ltd.).

The preferable form for investment by foreigners is either a limited liability company or a company limited by shares, regardless of whether a new company is to be founded or the foreigner would like to be a partner in an existing company. We shall hence discuss only these forms.

A limited liability company (LLC) is – according to Hungarian legislation – a company which is founded 'with a primary stock consisting of predetermined primary stakes, and other material contributions – if any – defined in the deed of association. The member is not liable for the obligations of the company.'[1]

The law prescribes what the deed of association should contain as a minimum: the amount of the primary stock and the primary stake, the way of its payment, the way of voting at the members' meeting, the first managers, their representation and the way of procuration, the first members of the supervisory board, if any, whether an auditor is required, the person of the first auditor, etc. The deed of association may also dispose about other issues, such as non-pecuniary contributions (in German: *Sacheinlage*), their characteristics and their value. The deed of association may also provide that other non-pecuniary contributions can be made by the members (subsidiary contributions) and lay down the conditions, including the case of non-performance.

1 The citations of Act VI of 1988 are from the English translation of this Act as published by the Information Service of the Hungarian Chamber of Commerce.

The deed of association may authorise the members' meeting to ask for additional contributions, may regulate the legal succession and the allowance of the withdrawal of business shares, etc. The Act regulates also the legal relationship between the company and its members, the organisation of the company, the members' meeting, the managers, the supervisory board, the amendment of the deed of association, the increase or reduction of the primary capital, and the termination of the company.

It is possible to found a one-man company. In such cases the deed of association is replaced by the deed of foundation.

The law provides that the primary capital of the LLC may not be less than 1 million HUF, while the value of the primary stakes may not be less than 100,000 HUF. A minimum of 30 per cent of the equity, but at least 500,000 HUF has to be paid at foundation, the rest within one year.

The members' meeting may increase or reduce the primary capital (but may not go below 1 million HUF). In case of increase of the primary capital the members have a right of preemption. (We shall later return to the nature of the non-pecuniary contribution.)

The supreme organ of the company is the members' meeting, the scope of authority of which is defined by the law. To its competence belongs the decision on practically every important matter except for the day-to-day business. It has to be convoked at least once a year. The administration of the business and the representation of the company is the duty of the managing director(s). Their task includes the keeping of the register of the members containing the equities, the subsidiary contributions and all other data on a day-to-day basis. Besides that, the managing directors are responsible for the company's keeping regular business files.

In case of an LLC it is obligatory to set up a supervisory board if the primary capital surpasses 20 million HUF, or if there are more than 25 members, or if the number of employees surpasses 200. The supervisory board must have at least 3 members, one of them being the chairman. The supervisory board is to be elected by the members' meeting.

A vote of at least 3/4 at the members' meeting is required for the amendment of the deed of association, while a unanimous resolution is needed for the increasing of the obligations of the members or the determination of new obligations, as well for the encroaching upon their rights. It is only possible to terminate the company with at least a 3/4 majority resolution. The law secures the protection of the creditors in case of the decrease of the primary capital or the termination of the company.

A *company limited by shares* (Co./Ltd.) is an economic association, which is 'formed with a registered capital consisting of shares of predetermined amount ... in which the members' [shareholders'] liability towards the company is limited to supplying the face value, or value of issue of the share.'

The shareholder has no liability for the obligations of the company. He is obliged to pay up the full value of the shares within one year from registration of the company.

The shareholder is entitled to a *pro rata* part of the profit shown in the balance sheet (the dividend). He may not claim back his non-pecuniary contribution during the existence nor in case of termination of the company. He may recover his invested capital (or a larger or smaller part thereof) through the transfer of his rights, i.e. by selling his shares (stocks). He has the right to participate at the general assembly and may ask questions there, may make remarks, and has the right in case of shares that carry a right to vote, to make proposals and to vote.

The general assembly is the supreme organ of the company. A number of matters belong to its exclusive competence, among others the adaption and amendment of the statutes, the increase and reduction of the registered capital, the merger, dissolution or termination of the company as well conversion into another company form. The executive organisation of the company is the board of directors and its control organ is the supervisory board, the members of which are elected by the general assembly for a predetermined period. It is obligatory that the company should elect an auditor and to use his services. There is also the possibility to render non-pecuniary contributions.

1.2. Stake in other companies

The Act on Economic Associations mentions in connection with companies limited by shares, that a company may achieve a considerable, a majority or a mutual interest in another company.

A *considerable interest* is achieved when a company owns more than 1/4 of shares of another company. We may speak of a *majority (controlling) interest* when a company has acquired shares representing more than 1/2 of the registered capital of another company (the controlled company). The members of the board of the company in possession of the majority interest in the controlled company may instruct the officers of the controlled company. These instructions, which have to be obeyed, are the vehicles of control.

A *mutual interest* exists when two or more companies are mutual proprietors of each other's shares.

In case of companies limited by shares the law contains certain restrictions on how great an interest a company may acquire in another company, and what the procedure for such acquisition is. There are however no such legislative limitations in case of other forms of associations. If a company acquires an interest of at least 25 per cent in another company, it is not yet an affiliation, because the 25 per cent stake does not mean *per se* the disposal over the company affected. This situation appears first in the case of a majority interest. Should the stake reach 75 per cent of the registered capital or shares of another company, this becomes a directly supervised company, a branch company.

Although the amendment of the Act of Economic Associations provides in § 321 that if a company limited by shares obtains a considerable, a majority or a mutual interest in another such company the regulations of the Unfair Market Behaviour Act (Act LXXXVI of 1990) will apply, the restrictions pertaining to licence and know-how agreements (dealt with in paragraph 4.3 of Chapter 3) are not stringent or only to a small extent. Foreign investors therefore prefer the formation of such joint ventures in case of technology transfers, giving them a considerable or possibly even a majority interest.

2. NON-PECUNIARY CONTRIBUTION (CONTRIBUTION IN KIND)

One may speak first of all in the case of limited liability companies and companies limited by shares of a non-pecuniary contribution. Paragraph (2) of § 22 of the Hungarian Act of Economic Associations says the following:

> 'The non-monetary contribution may consist of any negotiable thing having a property value, intellectual creation or valuable right (title).'

According to paragraph (3) of § 161 of the amended Act only such enforceable rights as well as intellectual creations or titles may be considered – even in case of the existence of the conditions mentioned in paragraph (2) of § 22 of the Act – which may be transferred to the company later without the permission of third persons. There is a further stipulation on non-monetary contributions in paragraph (2) of § 162 which says:

> 'Those members of the LLC who have knowingly caused the company to accept a valuation of a non-pecuniary stake of a member higher than its actual value ... shall bear joint and several liability for all damages arising therefrom. Exemption from such liability has no effect in respect of the creditors of the LLC.'

In case of a company limited by shares one may also speak about a non-monetary contribution. Paragraph (1) of § 253 of the amended text of the Act on Economic Associations stipulates in principle the same as paragraph (3) of § 161, quoted above, in connection with limited liability companies. Paragraph (2) of § 253 states that the founders are obliged to present those data in writing, on which the value of the non-pecuniary contribution was established, while paragraph (3) of this section says that in the founding draft the value of any non-pecuniary contribution may only be based at the value previously determined by an auditor.

The Act allows both in the case of a limited liability company and a company limited by shares that an increasing of the capital could be rendered in the form of a non-monetary contribution. In case of a company limited by shares this can be done even when the full countervalue of the shares was not put at the disposal of the company, i.e. the contribution to the registered capital has not been fully paid up.

It has to be mentioned that both the Swiss substantive law (*Obligationenrecht*) and the German *Handelsrecht* have very similar rules to the above. The Swiss law moreover is familiar with the concept of non-pecuniary contributions (in German: *Sacheinlage*) even in the case of a deposit partnership. Thus assets may be put at the disposal of economic associations (companies) in the form of a non-pecuniary contribution. From this it follows that the *intellectual creations or rights become the property of the company,* i.e. the company may later transfer them to others without the permission of the contributing member (shareholder). The salient point is that the contributor has to be responsible for the value of the intellectual property or title, i.e. that he has contributed the intellectual property or title at its real value to the company. This is the reason why possessors of such property or title will have to judge whether it should be contributed in this form, or rather through concluding a licence agreement.

For registered rights the assessment of value may be relatively easier, since some precedents are available for their evaluation. If non-registered rights, such as know-how, are the subject of the contribution, the setting of its fair price is almost impossible, leading to unquenchable altercations if the company later comes into difficulty. Therefore a contribution of this kind of intellectual property is always a doubtful issue. It is quite another issue however, when the company was founded precisely for the purpose of the utilisation of a given know-how or patent.

It has to be mentioned here, that the Economic Council of the Hungarian Courts has reached some decisions in respect of this matter. E.g. the non-pecuniary contribution of know-how was said to be a valid contribution according to one decision (GK 52/1989). It mentioned judged that although no exclusivity is attributed to know-how, it is still protected against unjustified use. The entitled may request, that the person acquiring or using it unjustly should share with him the financial gains. Hence know-how has a value and is negotiable intellectual property, and may be the subject of a non-pecuniary contribution.

The question arises how what extent the intellectual property rendered as a non-pecuniary contribution comes into the property of the company and to what extent it remains in the possession of its contributor. If the future contributor has already granted a licence to third parties prior to the contribution, one may wonder whether it is a valid contribution. These are open questions, the decision of which waits for the future.

The situation is much clearer in case of registered rights, e.g. a patent or a trade mark. In connection with the latter decision No GK 39/1992 of the Economic Council of the Courts says that 'the non-pecuniary contribution (namely that of a trade mark) according to paragraph (2) of § 161 of Act VI of 1988 did not take place, because *the trade mark has not been assigned as yet.*' The above decision states unequivocally that *registered* rights may only be contributed by an assignment. In the case of know-how this cannot be a precondition, since the know-how is not a registered intellectual property.

3. SUBSIDIARY CONTRIBUTION

Attention has to be drawn to the fact, that Swiss substantive law (*Obligationen-recht*) does not mention subsidiary contributions in connection with the foundation of companies. The German *Handelsgesetz* mentions it in the GmbHG section. (Care is advised since the texts from 1985 and 1991 are different!) A commentary of the older version[2] mentions that subsidiary contribution services may also be made by the cession of certain intellectual property (in German: *Schutzrechte*).

According to paragraph (2) of § 106 of the Hungarian Act on Economic Associations:

> 'The members of the *union* may also undertake a commitment to supply other valuable considerations [subsidiary contributions]. The member is entitled to an extra remuneration for the subsidiary contribution; such a remuneration has to be booked among the liabilities in the union's balance-sheet.'

Paragraph (2)(b) of § 128 of the same Act stipulates that in case of the foundation of *joint enterprises:*

> 'other valuable contributions [may] be supplied by the members [subsidiary contributions]; the terms thereof, as well as the amount of liquidated damages payable in case of failure, or partial failure, to supply such a subsidiary contribution,'

have to be dealt with in the deed of association. § 129 of the Act stipulates as follows:

> 'Besides supplying their material contributions, the members of the enterprise may undertake to provide other valuable contributions [subsidiary contributions]. The member is entitled to a special remuneration for such a subsidiary contribution; this shall be booked in the balance-sheet of the enterprise among the company's liabilities.'

A similar stipulation to the above is contained for limited liability companies in the Act (paragraph (2)(b) of § 157). Thus there is a possibility for the

2 Bambach-Hueck: *Kommentar zum GmbH Gesetz*, C.H. Beck'sche Verlagsbuchhandlung, München, 1985; comments to § 3 of the Act: remark 32 on p. 54 and 41 on p. 58.

members to effect a technology transfer as a subsidiary contribution. The advantage for the company of such contributions is that if the transfer does not happen in the form of a licence or know-how agreement, the member shall be liable for liquidated damages in respect of a delay of his contribution. Besides that – at least in our opinion – the member would have not only to contribute the basic know-how but also improvements thereof, which is not advantageous for the contributor. This is the reason why the transfer of technology in the form of a subsidiary contribution is only of limited importance and will only happen in special cases.

A further drawback in connection with the subsidiary contribution of transferable technologies is that if the member wishes to sell his share in accordance with the stipulations of the Act and with the deed of association, he will practically lose his subsidiary contribution.

Such subsidiary contributions are not advantageous for the company either. If registered rights have been contributed to the company and the contributor sells his shares, they may be maintained – depending on the deed of association –, but in case of non-registered rights there will be no transfer of later improvements. The joint venture may thus find itself in a problematic situation, which may perhaps be eliminated through complicated contracts only.

This is the reason why registered rights and technical or non-technical know-how are in most cases transferred neither as a non-pecuniary contribution nor as a subsidiary one, but within the context of a licence or know-how agreement concluded between the company and the member.

4. CONTRACTS IN RESPECT OF JOINT VENTURES

4.1. The foundation contract (deed of association)

When founding a company the parties have to agree on the terms of contract, and in the case of a company limited by shares they have to elaborate the statutes too. According to the Act on Economic Associations the following issues have to be determined in the deed of association:

- the trade name and seat of the company,
- the members (names or firms) and domiciles,
- the range of activity of the company,
- the size of the company's assets, when and how they must be made available,
- other details prescribed by the law for the various forms of association.

A typical example of a deed of association (foundation) of a limited liability company contains the following:

- the name of the founders,

- the deed of the founding (association),
- the name of the company in Hungarian and in foreign languages, possibly its abbreviation,
- the seat of the company,
- the amount of the primary capital (stock),
- the amount of the primary stakes per member, enumerating separately the pecuniary and the non-pecuniary contributions,
- the time and manner of rendering pecuniary and non-pecuniary contributions (e.g. half of the pecuniary contribution shall become due immediately, the other half only within one year, but the non-pecuniary contribution is to be provided to the company within 30 days of the foundation),
- the consequences of not meeting the above dates (e.g. after a warning an exclusion plus payment of damages),
- the rights and obligations of the members,
- the eventual subsidiary contributions (an effective participation in the activity),
- the duration of the company (in most cases undetermined),
- the organs of the company,
- the members' meeting and in connection with it the enumeration of those issues, which belong to its exclusive competence, the quorum of the members' meeting, the way of convocation (how many days beforehand it has to be convoked, with the simultaneous communication of the agenda),
- the eventual decision-making outside the members' meeting,
- the conducting of the members' meeting,
- the appointment of the first managing directors (and for how long), the manner of their election afterwards, tasks, obligations, rights, determination of their remuneration, incompatibility with other functions,
- the appointment of the members of the supervisory board (their tasks, obligations, rights, way of working, remuneration),
- the representation of the company, the rights of signature,
- the details of the transfer of a business share,
- the division of profits,
- the amendment of the deed of association,
- the increase of the primary capital,
- the reduction of the primary capital,
- the possible transformation, merger or association with other companies, and termination,
- the registration of the company,
- the governing law in respect of the deed of association, the jurisdiction in cases of litigation,
- possible other regulations.

Joint ventures are often founded in the form of limited liability companies. Companies limited by shares are to be founded in a more binding way, so in the statutes of such a company there are even more issues to be regulated than in the deed of association of a limited liability company. *In lieu* of the members' meeting there is a general assembly and instead of primary capital there is registered capital, and the business stakes are substituted by shares (stocks)./

The value of the non-pecuniary contribution, – especially if it is an intellectual property or right – is a very delicate issue in the case of both types of companies, and it has to be defined precisely in the deed of association or the statutes. The same relates to possible subsidiary contributions.

4.2. The syndicate contract

All other issues which have nothing to do with the formation of the company but pertain to its operation are to be regulated in a syndicate contract. This may be very varied indeed. Below we will show an outline of a possible contract:

- *The parties.*

- *The subject of the contract.*
  In this chapter of the contract it should be put down, what the purpose of the cooperation of the parties is and the aim of founding a joint venture (e.g. common production based on the licence of the foreign partner, sale of the product maybe including its export, the supply of certain raw materials, technical assistance, etc.). If applicable:
  · the licence, know-how or other knowledge, which has to be rendered by the foreign partner (it is practical to conclude in respect of the licence or know-how a separate agreement. This has to be indicated in the syndicate contract, or such an agreement might even form an appendix to the syndicate contract);
  · the transfer of the commercial know-how. Should this be an important one, a separate agreement is to be concluded for this purpose too;
  · the trade mark licence and its conditions (if necessary a separate agreement has to be concluded);
  · an indication of the engineering and/or technical assistance (here also a separate agreement is to be concluded if necessary).

- *The obligations of the partners*, such as
  · a detailed description of the factory or unit and of the machines to be put at the disposal of the joint venture by both parties, the machinery to be delivered by the foreign partner (transferor) as well as its source and procurement, the delivery of certain raw materials, etc.;
  · market agreements (e.g. that the domestic partner will take over products from the foreign partner in order to start the market penetration until the production of the joint venture is in full swing; or the obligation of the foreign partner to take over a certain amount of the product);

· the enumeration of those agreements which shall complete the syndicate agreement but are to be concluded later.

- *The characteristics of the joint venture.*
  It has to be stipulated here which form the joint venture company shall have (limited liability company, company limited by shares, etc.), as well as its seat and name. All matters which could not be laid down in the deed of association have to be regulated here; e.g. by whom and how the directors and members of the supervisory board will be nominated. The monetary and the non-pecuniary contribution of the parties shall be determined here, as well as the possibility of concluding extra agreements on this matter (e.g. the possible increase of the primary stock or of the registered capital).

- *Failure of the parties.*
  Here should be determined what happens if one of the parties does not fulfil its obligations for putting the promised goods or services at the disposal of the joint venture or does it only partially, or does not or only partially fulfil its possible later obligations in respect of the increase of the primary stock or of the registered capital, etc.

- *Market policy.*
  In this chapter the market cooperation has to be regulated, further if the parties effect their sales directly or through another company.

- *Profit policy.*
  The parties have to clarify what should happen with the eventual profit, what part of it shall be the dividend, what part is go to investment and what is the mechanism of decision-making in this respect.

- *The eventual assignment of the business shares or stocks.*
  Should one of the parties wish to sell its business shares or stocks, does the other party buy it, or should it be sold to third parties and under which procedure?

- *General legal dispositions.*
  The issues of *force majeure* (Act of God), the governing law, and the procedure in case of litigation (arbitration court) should be discussed under this heading.

- *Coming into force.*
  The coming into force may not be trivial. It is possible that a large number of contracts is drawn up, to be executed one by one, including the deed of association. It seems to be a practical solution, in case some of the agreements cannot be concluded simultaneously (such as the engineering or the technical assistance agreement), that a deadline should be fixed for their

conclusion. The execution of the deed of association and the syndicate contract however cannot be postponed.

4.3. Licence and know-how agreements

As seen above, there are three ways to bring in technical knowledge or non-technical know-how and registered rights (e.g. a patent, trade mark, etc.) into a joint venture: the non-pecuniary contribution, the subsidiary contribution and a separate licence and/or know-how agreement to be concluded *simultaneously* with the deed of association and the syndicate agreement. A non-pecuniary contribution comes to the fore if one of the parties (members) cannot or does not want to contribute in cash, but this is relatively rare. The parties usually prefer to conclude a licence or know-how agreement, since otherwise the patent, trade mark or know-how to be transferred becomes practically the property of the joint venture company. Moreover a separate licence and/or know-how agreement can more easily be terminated under the conditions laid down in the contract, in case the parties are unable to cooperate smoothly within the joint venture.

The subsidiary contribution may be advantageous for the domestic party, but it is not advantageous for the foreign party, or only in special cases. Such companies are hence rare where the technology is transferred in the form of a subsidiary contribution.

5. ANTI-TRUST RULES AND JOINT VENTURES

The amended version of § 321 of the Act on Economic Associations states that in case of a considerable or majority interest in a company limited by shares, the rules of the Act on the Prohibition of Unfair Market Behaviour are to be followed. It is unclear how this may affect the possibility to limit the freedom of contract (*see* paragraph 4.3 of Chapter 3) in the case of a joint venture, especially an affiliated or subsidiary company. The problem is not specific to Hungary but generally arises where anti-trust legislation exists.

If technical or non-technical knowledge or exclusive rights are transferred to the joint venture in the form of a non-pecuniary or subsidiary contribution, the licence aspects of the antitrust regulations may be totally disregarded. Therefore it has to be investigated only what happens in the case of separate licence and/or know-how agreements.

In the case of licences the 'tie-in' prohibition mentioned above in paragraph 4.3 of Chapter 3 – i.e. the prohibition of obliging the licensee to purchase non-patented goods – will probably not prevail, since an affiliated or subsidiary company may buy various products from the parent company. The so-called 'tie-out' provision – i.e. the prohibition of the sale of similar goods of the competition – needs no special contractual stipulation, since the licensor has such a big stake in the affiliated or subsidiary company that he may frustrate such sales.

The prohibition of the abuse of 'package licensing' may not prevail either, since the overwhelming interest of the affiliated or subsidiary company is that the parent company should put at its disposal as much as possible of its licences and/or know-how. Basing the licence fee on the full turnover is also natural, at least in case of a majority interest.

The prohibition of the collecting of licence fees after the expiry of a patent shall also not prevail, since the parent company will declare it as a mixed licence (patent plus know-how) if this is more favourable than payment of mere dividends.

The prohibition of the non-impugnment of the patent is of no importance in the case of an affiliated or subsidiary company, since the interest of the latter is to take no action against the patent.

Among the discretionary clauses the issue of market limitation of the licensee probably loses its importance, since the cooperation between the parent company and the affiliated or subsidiary joint venture company is usually connected to a meaningful market sharing. The competition between them is very small and in practice nonexistent. For this reason also the so-called 'grant-back' clauses are of minor importance, since the partners cooperate with each other very closely, making such stipulations superfluous.

Briefly: in the relation between a parent company and its affiliated or subsidiary company the topics discussed under the heading of limitation of the freedom of contracting are practically meaningless, since the limitation is asserted *per se* through the special dependence-cooperation relationships. This may be one of the reasons, why for the transfer of technology the establishment of a joint venture is preferred by many.

Literature

A. In Hungarian

Gazda, I., *Az ipari know-how* (The industrial know-how), OMKDK, Budapest, 1969.
Gazda/Kövesdi Vida: *Találmányok, szabadalmak. Mûszaki alkotások jogvédelme és értékesítése* (Inventions, patents. The protection and sale of technical creations), Közgazdasági és Jogi Könyvkiadó, Budapest, 1985.
CHEMOKOMPLEX, *Külkereskedelmi szerződéskötési ismeretek I-II* (Contracting practices in foreign trade), published by the foreign trade company CHEMOKOMPLEX, manuscript, Budapest, 1979.
Lontai, E., *Licenciaszerződések alapvető kérdései* (The basic issues of licensing agreements), Közgazdasági és Jogi Könyvkiadó, Budapest, 1978.

B. In English

Al Ali, S., 'Laws and regulations on technology transfer to developing countries,' *Sci-Publ. Pol.* (Guilford) 1991, No. 5, p. 295–300.
Contractor, F.J., *Licensing in International Strategy: A Guide for Planning and Negotiations*, Quorum Books, Westport, Conn., 1985.
Dessemontet, F., *The Legal Protection of Know-how in the United States of America*, Droz Geneva, Rothman, South Kackensack, New Jersey, 1976.
'Digging out of ruins: Technology transfer,' *Nature* (London), 20 February 1992, p. 671.
Gazda, I., 'Technical guarantees in licence and know-how agreements,' *Proc. of the Hungarian IAPIP Group* No. 3, 1973, p. 35–40.
— *Summary of the Section Know-how*, Conference of the Hungarian Group of IAPIP, Budapest, 24–28 September 1973, Hung. Central Technical Library and Documentation Centre, Budapest, 1974.
— *Summary of the Section Technology Transfer and Industrial Property Protection*, Conference of the Hungarian Group of IAPIP, Budapest, 13–17 September 1982, OMIKK-Technoinform, Budapest, 1983.
— 'Know-how as it should be,' *Proc. of the Hungarian IAPIP Group*, No. 11–12, 1983, p. 44–52.
— 'Technology transfer between Market Economy and former socialist countries,' *Proc. of the Hungarian AIPIP Group*, No. 18, 1991, p. 69–80.

Gedeon, S., 'On the problem of how to define the invention and the know-how,' *Proc. of the Hungarian IAPIP Group*, No. 10, 1980, p. 23–34.
Hall, P., *Technology Innovations and Economy Policy*, Allan, VIII, Oxford, 1988.
Mádi, Cs., 'Transfer of technology. Hungary in the eighties. Hungary's trade in the intellectual products,' *Acta Oecon.*, 1990, No. 1–2. p. 141–168.
Marer, F., *East-West Technology Transfer. Study of Hungary*, 1968–84. OECD, Paris. 1986.
Moore, J.H., 'Knowledge transfer in the United States,' *Europ. Econ. R.*, Amsterdam, 1988, 2–3, p. 581–603.
Polzien and Langen, *International Licensing Agreements*, Second Edition, The Bobbs-Merrill Company Inc., 1973.
Third International Civil Law Conference in Stockholm on Technology Transfer, *Conference Papers* (manuscript), 10–14 June 1991.

C. In German

Lüdecke, W. and Fischer, E., *Lizenzverträge*, Verlag Chemie, GmbH, Weinheim, Bergstr., 1957.
Stumpf, H., *Der know-how Vertrag*, Verlagges. Recht und Wirtschaft mbH, Heidelberg, 1970.
Vida, A., *Das Warenzeichen in der Wirtschaft der socialistischen Länder*, C. Heymans Verlag, Köln, Berlin, Bonn, München, 1992.

Appendix

I. US ANTITRUST STATUTES

A. Section 1 of the Sherman Act (15 U.S.C. § 1).

Every contract, combination in the form of trust or otherwise, or conspiracy, in restraint of trade or commerce among the several States, or with foreign nations, is declared to be illegal.

B. Section 2 of the Sherman Act (15 U.S.C. § 2).

Every person who shall monopolize, or attempt to monopolize, or combine or conspire with any other person or persons, to monopolize any part of the trade or commerce among the several States, or with foreign nations, shall be deemed guilty of a misdemeanor ....

C. Section 3 of the Clayton Act (15 U.S.C. § 14).

It shall be unlawful for any person engaged in commerce, in the course of such commerce, to lease or make a sale or contract for sale of goods, wares, merchandise, machinery, supplies, or other commodities, whether patented or unpatented, for use, consumption or resale within the United States ... or fix a price charged therefor, or discount from, or rebate upon, such price, on the condition, agreement, or understanding that the lessee or purchaser thereof shall not use or deal in the goods, wares, merchandise, machinery, supplies, or other commodities of a competitor or competitors of the lessor or seller, where the effect of such lease, sale, or contract for sale or such condition, agreement, or understanding may be to substantially lessen competition or tend to create a monopoly in any line of commerce.

D. Section 7 of the Clayton Act (15 U.S.C. § 18).

No corporation engaged in commerce shall acquire, directly or undirectly, the whole or any part of the stock or other share capital and no corporation subject to the jurisdiction of the Federal Trade Commission shall acquire the whole or any part of the assets of another corporation engaged also in commerce, where in any line of commerce in any section of the country, the effect of such acquisition may be substantially to lessen competition, or to tend to create a monopoly.

E. *Section 5(a) (1) of the Federal Trade Commission Act (15 U.S.C. § 45(a) (1)*

Unfair methods of competition in commerce, and unfair or deceptive acts or practices in commerce, are declared unlawful.

II. ROME TREATY

Article 85

1. The following shall be prohibited as incompatible with the common market: all agreements between undertakings, decisions by associations of undertakings and concerted practices which may affect trade between Member States and which have as their object or effect the prevention, restriction or distortion of competition within the common market, and in particular those which:
(a) directly or indirectly fix purchase or selling prices or any other trading conditions:
(b) limit or control production, markets, technical development, or investment:
(c) share markets or sources of supply:
(d) apply dissimilar conditions to equivalent transactions with other trading parties, thereby placing them at a competitive disadvantage:
(e) make the conclusion of contracts subject to acceptance by the other parties of supplementary obligations which, by their nature or according to commercial usage, have no connection with the subject of such contracts.

2. Any agreements or decisions prohibited pursuant to this Article shall be automatically void.

3. The provisions of paragraph 1 may, however, be declared inapplicable in the case of:
- any agreement or category of agreements between undertakings:
- any decision or category of decisions by associations of undertakings:
- any concerted practice or category of concerted practices:

which contributes to improving the production or distribution of goods or to promoting technical or economic progress, while allowing consumers a fair share of the resulting benefit, and which does not:

(a) impose on the undertakings concerned restrictions which are not indispensable to the attainment of these objectives:
(b) afford such undertakings the possibility of eliminating competition in respect of a substantial part of the products in question.

Article 86

Any abuse by one or more undertakings of a dominant position within the common market or in a substantial part of it shall be prohibited as incompatible with the common market in so far as it may affect trade between Member States.

Such abuse may, in particular, consist in:
(a) directly or indirectly imposing unfair purchase or selling prices or other unfair trading conditions:
(b) limiting production, markets or technical development to the prejudice of consumers:
(c) applying dissimilar conditions to equivalent transactions with other trading parties, thereby placing them at a competitive disadvantage:
(d) making the conclusion of contracts subject to acceptance by the other parties of supplementary obligations which, by their nature or according to commercial usage, have no connection with the subject of such contracts.